6th Symposium on Large TPCs for Low Energy Rare Event Detection 2012

Journal of Physics: Conference Series Volume 460

Paris, France
17 - 19 December 2012

ISBN: 978-1-62993-935-3
ISSN: 1742-6588

Printed from e-media with permission by:

Curran Associates, Inc.
57 Morehouse Lane
Red Hook, NY 12571

Some format issues inherent in the e-media version may also appear in this print version.

Printed by Curran Associates, Inc. (2014)

For permission requests, please contact the Institute of Physics at the address below.

Institute of Physics
Dirac House, Temple Back
Bristol BS1 6BE UK

Phone: 44 1 17 929 7481
Fax: 44 1 17 920 0979

techtracking@iop.org

Additional copies of this publication are available from:

Curran Associates, Inc.
57 Morehouse Lane
Red Hook, NY 12571 USA
Phone: 845-758-0400
Fax: 845-758-2634
Email: curran@proceedings.com
Web: www.proceedings.com

TABLE OF CONTENTS

Preface

For the sixth time the International Symposium on large TPCs for Low-Energy Rare-Event Detection has been organized in Paris on 17-19 December 2012. As for the previous conference, we were welcomed in the Astroparticle and Cosmology Laboratory (APC).

Around one hundred physicists from all over the world gathered to discuss progress in the dark matter and low-energy neutrino search. The new results from the LHC were also widely discussed. The Higgs discovery at 125 GeV, without any sign of other new heavy particles, does not provide us with any information on the nature of dark mater.

Alternatives to the favored SUSY model, in which the role of the WIMP is played by a stable neutralino, predict low mass candidates below a few GeV. Developing low threshold detectors at sub-keV energies becomes mandatory, and interest for Axion or Axion-like particles as dark matter is revived.

We have seen increasing activity in the field and new infrastructures for these searches have been developed. We heard news of activities in the Canfranc laboratory in Spain, Jinping in China, SURF in the USA and about the extension project of Fréjus (LSM) laboratory. We would like to thank the organizing and advisory committees as well as the session chairpersons: J Zinn-Justin, G Wormser, D Nygren, G Chardin, F Vannucci, D Attié, T Patzak and S Jullian.

I Giomataris, P Colas, I G Irastorza

Group Picture

Journal of Physics: Conference Series **460** (2013) 011002 doi:10.1088/1742-6596/460/1/011002

Peer review statement

All papers published in this volume of *Journal of Physics: Conference Series* have been peer reviewed through processes administered by the proceedings Editors. Reviews were conducted by expert referees to the professional and scientific standards expected of a proceedings journal published by IOP Publishing.

Journal of Physics: Conference Series **460** (2013) 012001 doi:10.1088/1742-6596/460/1/012001

Search for Dark Matter at the LHC

Mario Martínez

ICREA/Institut de Fìsica d'Altes Energies, Barcelona 08193, Spain

E-mail: mmp@ifae.es

Abstract. In this contribution, recent results on the search for dark matter candidates at the LHC in monojet and monophoton final states are presented based on the data collected by the ATLAS and CMS experiments at $\sqrt{s} = 7$ TeV.

1. Introduction

Events with an energetic jet or photon and large missing momentum in the final state constitute a clean and distinctive signature in searches for new physics at colliders. In particular, monophoton and monojet final states have been studied [1, 2, 3, 4, 5, 6, 7, 8, 9] in the context of searches for supersymmetry and large extra spatial dimensions (LED), aiming to provide a solution to the mass hierarchy problem, and the search for weakly interacting massive particles (WIMPs) as candidates for dark matter (DM).

The presence of a non-baryonic DM component in the universe is inferred from the observation of its gravitational interactions [10], although its nature is otherwise unknown. A WIMP χ with mass m_χ in the range between 1 GeV and a few TeV is a plausible candidate for DM. It could be detected via its scattering with heavy nuclei [11], the detection of cosmic rays (energetic photons, electrons, positrons, protons, antiprotons, or neutrinos) from $\chi\bar{\chi}$ annihilation in astrophysical sources [10], or via $\chi\bar{\chi}$ pair-production at colliders where the WIMPs do not interact with the detector and the event is identified by the presence of an energetic photon or jet from initial-state radiation. The interaction of WIMPs with Standard Model (SM) particles is assumed to be driven by a mediator with mass at the TeV scale and described using a non-renormalizable effective theory [12] with several operators. The vertex coupling is suppressed by an effective cut-off mass scale $M_* \sim M/\sqrt{g_1 g_2}$, where M denotes the mass of the mediator and g_1 and g_2 are the couplings of the mediator to the WIMP and SM particles.

2. Results

Both the ATLAS and CMS collaborations have carried out measurements on monojet and monophoton production in proton-proton collisions at $\sqrt{s} = 7$ TeV[1]. In all cases, good agreement is observed between the data and the standard model (SM) predictions. 90% confidence level (CL) upper limits on the pair production cross section of dark matter WIMP candidates are determined, and then translated into upper limits on the nucleon-WIMP interaction cross section using the prescription in Refs. [12, 13]. Figure 1 shows 90% CL upper limits on the nucleon-WIMP cross section as a function of m_χ for both monojet and monophoton results. The

[1] At the time of preparing this proceedings 8 TeV results are becoming available.

exclusion in the region 1 GeV $< m_\chi <$ 3.5 GeV (1 GeV $< m_\chi <$ 1 TeV) for spin-independent (spin-dependent) nucleon-WIMP interactions is driven by the results from collider experiments, with the assumption of the validity of the effective theory employed.

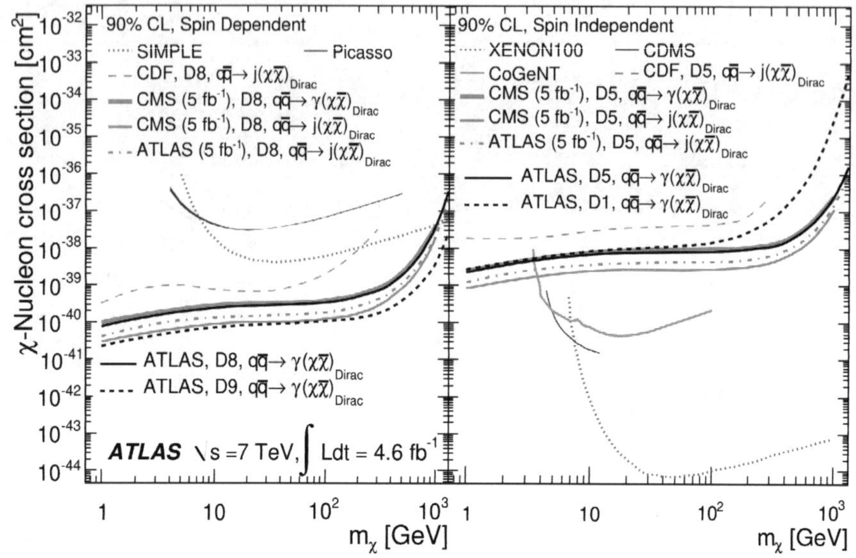

Figure 1. 90% CL upper limits on the nucleon-WIMP cross section as a function of m_χ for spin-dependent (left) and spin-independent (right) interactions [12, 13]. It includes monojet and monophoton results at colliders [4, 5, 6, 8, 9] compared to results from direct detection experiments [11].

References

[1] OPAL Collaboration, Eur. Phys. J. C **18**, 253 (2000); ALEPH Collaboration,Eur. Phys. J. C **28**, 1 (2003); L3 Collaboration, Phys. Lett. B **587**, 16 (2004); DELPHI Collaboration, Eur. Phys. J. C **38**, 395 (2005).
[2] D0 Collaboration, Phys. Rev. Lett. **101**, 011601 (2008).
[3] CDF Collaboration, Phys. Rev. Lett. **101**, 181602 (2008).
[4] CDF Collaboration, Phys. Rev. Lett. **108**, 211804 (2012).
[5] CMS Collaboration, Phys. Rev. Lett. **107**, 201804 (2011); CMS Collaboration, JHEP **09**, 094 (2012).
[6] CMS Collaboration, Phys. Rev. Lett. **108**, 261803 (2012).
[7] ATLAS Collaboration, Phys. Lett. B **705**, 294 (2011).
[8] ATLAS Collaboration, JHEP **04**, 075 (2013).
[9] ATLAS Collaboration, Phys. Rev. Lett. **110**, 011802 (2013).
[10] G. Bertone, D. Hooper, and J. Silk, Phys. Rep. **405**, 279 (2005).
[11] XENON100 Collaboration, Phys. Rev. Lett. **107**, 131302 (2011); CDMS Collaboration, Phys. Rev. Lett. **106**, 131302 (2011); CoGENT Collaboration, Phys. Rev. Lett. **106**, 131301 (2011); M. Felizardo *et al.*, arXiv:1106.3014 (2011); PICASSO Collaboration, Phys. Lett. B **711**, 153 (2012).
[12] J. Goodman *et al.*, Phys. Rev D **82**, 116010 (2010).
[13] In consultation with the authors of Ref.[12], a factor 4.7×10^{-39}cm^2 is used in the cross section formula for D8 and D9 operators instead of the quoted 9.18×10^{-40}cm^2.

Journal of Physics: Conference Series **460** (2013) 012002 doi:10.1088/1742-6596/460/1/012002

Future axion searches with the International Axion Observatory (IAXO)

I. G. Irastorza[1], F. T. Avignone[2], G. Cantatore[3], J. M. Carmona[1], S. Caspi[4], S. A. Cetin[5], F. E. Christensen[6], A. Dael[7], T. Dafni[1], M. Davenport[4], A.V. Derbin[8], K. Desch[9], A. Diago[1], B. Döbrich[23], A. Dudarev[4], C. Eleftheriadis[10], G. Fanourakis[11], E. Ferrer-Ribas[7], J. Galán[7], J. A. García[1], J. G. Garza[1], T. Geralis[11], B. Gimeno[12], I. Giomataris[7], S. Gninenko[13], H. Gómez[1],E. Guendelman[14], C. J. Hailey[15], T. Hiramatsu[16], D. H. H. Hoffmann[17], D. Horns[18], F. J. Iguaz[1], J. Isern[19], A. C. Jakobsen[6], J. Jaeckel[20], K. Jakovčić[21], J. Kaminski[9], M. Kawasaki[22], M. Krčmar[21], C. Krieger[9], B. Lakić[21], A. Lindner[23], A. Liolios[10], G. Luzón[1], I. Ortega[1], T. Papaevangelou[7], M. J. Pivovaroff[24], G. Raffelt[25], J. Redondo[25], A. Ringwald[23], S. Russenschuck[4], J. Ruz[24], K. Saikawa[22], I. Savvidis[10], T. Sekiguchi[22], I. Shilon[4], P. Sikivie[26], H. Silva[4], H. ten Kate[4], A. Tomas[1], S. Troitsky[13], T. Vafeiadis[4], K. van Bibber[27], P. Vedrine[7], J. A. Villar[1], J. K. Vogel[24], L. Walckiers[4], W. Wester[28], S. C. Yildiz[5], K. Zioutas[29]

[1]Laboratorio de Física Nuclear y Altas Energías, Universidad de Zaragoza, Zaragoza, Spain
[2]Physics Department, University of South Carolina, Columbia, SC, USA
[3]Instituto Nazionale di Fisica Nucleare (INFN), Sezione di Trieste and Università di Trieste, Trieste, Italy
[4]European Organization for Nuclear Research (CERN), Genève, Switzerland
[5]Dogus University, Istanbul, Turkey
[6]Technical University of Denmark, DTU Space Kgs. Lyngby, Denmark
[7]IRFU, Centre d'Études Nucléaires de Saclay (CEA-Saclay), Gif-sur-Yvette, France
[8]St.Petersburg Nuclear Physics Institute, St.Petersburg, Russia
[9]Physikalisches Institut der Universität Bonn, Bonn, Germany
[10]Aristotle University of Thessaloniki, Thessaloniki, Greece
[11]National Center for Scientific Research Demokritos, Athens, Greece
[12]Instituto de Ciencias de las Materiales, Universidad de Valencia, Valencia, Spain
[13]Institute for Nuclear Research (INR), Russian Academy of Sciences, Moscow, Russia
[14]Physics department, Ben Gurion Uiversity, Beer Sheva, Israel
[15]Columbia Astrophysics Laboratory, New York, USA
[16]Yukawa Institute for Theoretical Physics, Kyoto University, Kyoto, Japan
[17]Technische Universität Darmstadt, IKP, Darmstadt, Germany
[18]Institut für Experimentalphysik, Universität Hamburg, 22761 Hamburg, Germany
[19]Institut de Ciències de l'Espai (CSIC-IEEC), Facultat de Ciències, Campus UAB, Bellaterra, Spain
[20]Institut für theoretische Physik, Universität Heidelberg, Philosophenweg 16, 69120 Heidelberg, Germany
[21]Rudjer Bošković Institute, Zagreb, Croatia
[22]Institute for Cosmic Ray Research, University of Tokyo, Tokyo, Japan
[23]Deutsches Elektronen-Synchrotron DESY, Hamburg, Germany

[24]Lawrence Livermore National Laboratory, Livermore, CA, USA
[25]Max-Planck-Institut für Physik, Munich, Germany
[26]Department of Physics, University of Florida, Gainesville, FL 32611, USA
[27]Department of Nuclear Engineering, University of California Berkeley, Berkeley, CA, USA
[28]Fermi National Accelerator Laboratory, Batavia, IL, USA
[29]Physics Department, University of Patras, Patras, Greece

E-mail: Igor.Irastorza@cern.ch

Abstract. The International Axion Observatory (IAXO) is a new generation axion helioscope aiming at a sensitivity to the axion-photon coupling of $g_{a\gamma} \gtrsim$ few$\times 10^{-12}$ GeV^{-1}, i.e. 1-1.5 orders of magnitude beyond the one achieved by CAST, currently the most sensitive axion helioscope. The main elements of IAXO are an increased magnetic field volume together with extensive use of x-ray focusing optics and low background detectors, innovations already successfully tested in CAST. Additional physics cases of IAXO could include the detection of electron-coupled axions invoked to explain the white dwarf cooling, relic axions, and a large variety of more generic axion-like particles (ALPs) and other novel excitations at the low-energy frontier of elementary particle physics.

1. Introduction

The Peccei-Quinn (PQ) mechanism of dynamical symmetry restoration [1, 2] stands out as the most compelling solution of the strong CP problem. Central to the PQ mechanism is the axion [3, 4], the Nambu-Goldstone boson of a new spontaneously broken symmetry $U(1)_{PQ}$. The properties of axions allow them to be produced in the early universe as coherent field oscillations and as such to provide all or part of the cold dark matter [5, 6].

It is still possible to find these "invisible axions" in realistic search experiments and in this way test a fundamental aspect of QCD. The generic $a\gamma\gamma$ vertex allows for axion-photon conversion in external electric or magnetic fields in analogy to the Primakoff effect for neutral pions. As shown in 1983 by Pierre Sikivie, the smallness of the axion mass allows this conversion to take place coherently over macroscopic distances, compensating for the smallness of the interaction strength [7]. Especially promising is to use the Sun as a source for axions produced in its interior by the Primakoff effect. Directing a strong dipole magnet toward the Sun allows one to search for keV-range x-rays produced by axion-photon conversion, a process best visualized as a particle oscillation phenomenon [8] in analogy to neutrino flavor oscillations. Three such helioscopes have been built, in Brookhaven [9], Tokyo [10] and at CERN [11]. The CERN Axion Solar Telescope (CAST) has just finished a 8-year long data taking period, having strongly improved on previous experiments and even surpassed astrophysical limits in some range of parameters, although axions have not been found.

We have shown [12] that large improvements in magnetic field volume, x-ray focusing optics and detector backgrounds with respect to CAST are possible. Based on these improvements, and on the experience gathered within CAST, we propose the International Axion Observatory (IAXO), a new generation axion helioscope. IAXO could search for axions that are 1–1.5 orders of magnitude more weakly interacting than those allowed by current CAST constraints. It appears conceivable to surpass the SN 1987A constraint on the axion mass, $m_a \lesssim 20$ meV, test the white-dwarf (WD) cooling hypothesis [13], and explore a substantial part of uncharted axion territory experimentally. Moreover, IAXO would explore other more generic models of weakly interacting sub-eV particles (WISPs) [14, 15], in particular some models for axion-like particles (ALPs) that have been invoked in the context of several unexplained astrophysical observations. Equipped with microwave cavities or antennas, this setup could also aim at detecting relic axions [16, 17, 18].

Journal of Physics: Conference Series **460** (2013) 012002 doi:10.1088/1742-6596/460/1/012002

2. Experimental setup and expected sensitivity

IAXO will follow the basic conceptual layout of an enhanced axion helioscope seen in figure 1, implemented to a toroidal design for the magnet, together with x-ray optics and detectors attached to each of the magnet bores. The improvements anticipated for each of the experimental parameters of the helioscope were quantified in [12], organized in four scenarios (IAXO 1 to 4) ranging from most conservative to most optimistic values (see table 1 of [12]). These values are justified by several considerations of the magnet, x-ray optics and detectors, that are briefly outlined in the following, but we refer to [12] for a detailed discussion.

The magnet parameters are the ones contributing mostly to the helioscope's figure of merit. The CAST success has relied, to a large extent, on the availability of the first class LHC test magnet which was recycled to become part of the CAST helioscope. While going beyond CAST magnet's B or L is difficult, the improvement may come however in the cross section area, which in the case of the CAST magnet is only 3×10^{-3} m^2. Substantially larger cross sections can be achieved, although one needs a different magnet configuration. It is an essential part of our proposal that a new magnet must be designed and built specifically for this application, if one aims at a substantial step forward in sensitivity. A toroidal configuration for the IAXO magnet is being studied with a total cross section area A of up to few m^2, while keeping the product of BL close to levels achieved for CAST [19].

Another area for improvement will be the x-ray optics. Although CAST has proven the concept, only one of the four CAST magnet bores is equipped with optics. The use of focusing power in the entire magnet cross section A is implicit in the figures of merit defined in [12], and therefore the improvement obtained by enlarging A comes in part because a correspondingly large optic is coupled to the magnet. The optics challenge in this case is two-fold: not only must we optimize the optic figure of merit, but we must also consider the availability of cost-effective x-ray optics of the required size. IAXO's optics specifications can be met by a dedicated fabrication effort based on segmented glass substrate optics like the ones of HEFT or NuSTAR [20].

Finally, CAST has enjoyed the sustained development of its detectors towards lower

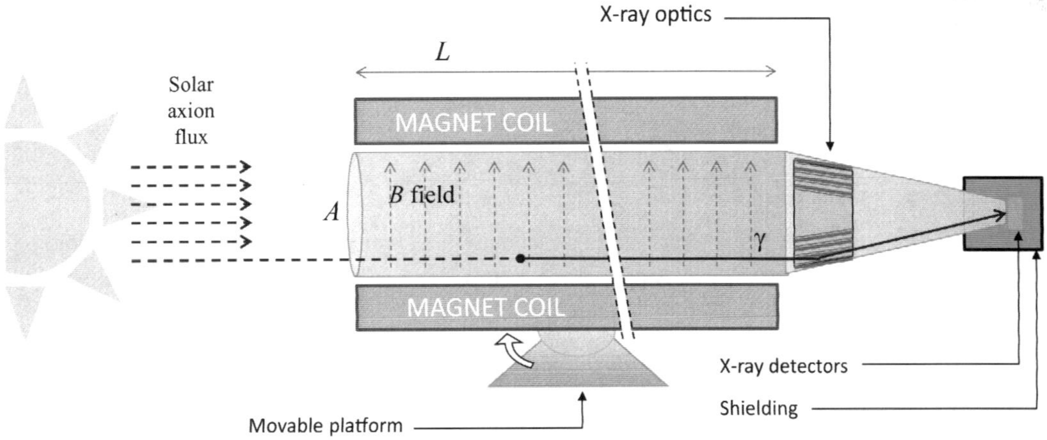

Figure 1. Conceptual arrangement of an enhanced axion helioscope with x-ray focalization. Solar axions are converted into photons by the transverse magnetic field inside the bore of a powerful magnet. The resulting quasi-parallel beam of photons of cross sectional area A is concentrated by an appropriate x-ray optics into a small spot area a in a low background detector. The envisaged design for IAXO includes eight such magnet bores, with their respective optics and detectors.

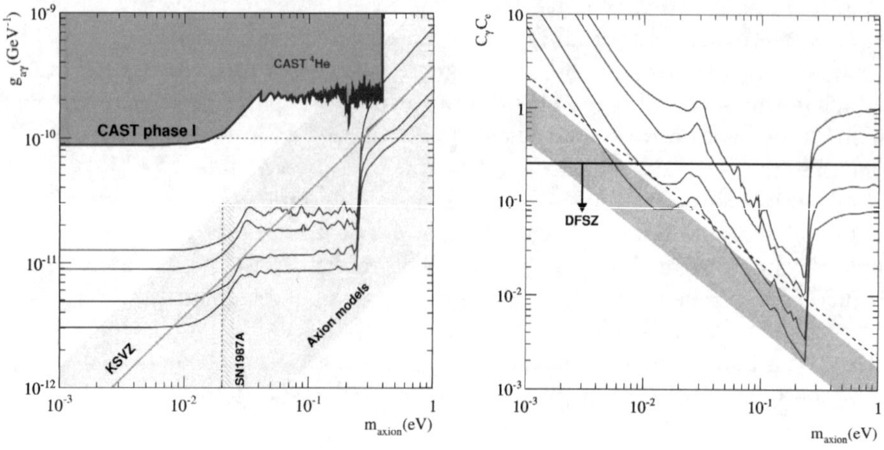

Figure 2. LEFT: The parameter space for hadronic axions and ALPs. The CAST limit, some other limits, and the range of PQ models (yellow band) are also shown. The blue lines indicate the sensitivity of the four scenarios discussed in the text. RIGHT: The expected sensitivity regions of the same four scenarios in the parameter space of non-hadronic axions with both electron and photon coupling. The orange band represents the region motivated by WD cooling, and the dashed line along the diagonal the red giants bound on the electron coupling. See [12] for details.

backgrounds during its lifetime. The latest generation of Micromegas detectors in CAST are achieving backgrounds of $\sim 5 \times 10^{-6}$ counts keV^{-1} cm^{-2} s^{-1}. This value is already a factor 20 better than the backgrounds recorded during the first data-taking periods of CAST. Prospects for reducing this level to 10^{-7} counts keV^{-1} cm^{-2} s^{-1} or even lower appear feasible [21, 22].

The computed sensitivities of each of the four IAXO scenarios are represented by the family of blue lines in figure 2, both for hadronic axions (left) and non-hadronic ones (right). They include two data taking campaigns for each of the scenarios: one three years long performed without buffer gas (analogous to CAST I), and another three years long period with varying amounts of ^{4}He gas inside the magnet bore (analogous to CAST II, although without the need to use ^{3}He). In general, IAXO sensitivity lines go well beyond current CAST sensitivity for hadronic axions and progressively penetrate into the decade 10^{-11}–10^{-12} GeV^{-1}, with the best one approaching 10^{-12} GeV^{-1}. They are sensitive to realistic QCD axion models at the 10 meV scale and exclude a good fraction of them above this. For non-hadronic axions, IAXO sensitivity lines penetrate in the DFSZ model region, approaching or even surpassing the red-giant constraints. Most relevantly, the IAXO 3 and IAXO 4 scenarios start probing the region of parameter space highlighted by the cooling of WDs.

Acknowledgements

We acknowledge support from the Spanish Ministry of Economy and Competitiveness (MINECO) under contract FPA2008-03456 and FPA2011-24058, as well as under the CPAN project CSD2007-00042 from the Consolider-Ingenio2010 program. Part of these grants are funded by the European Regional Development Fund (ERDF/FEDER). We also acknowledge support from the European Commission under the European Research Council T-REX Starting Grant ERC-2009-StG-240054 of the IDEAS program of the 7th EU Framework Program. Part of this work was performed under the auspices of the U.S. Department of Energy by Lawrence Livermore National Laboratory under Contract DE-AC52-07NA27344 with support from the

LDRD program through grant 10–SI–015. Partial support by the Deutsche Forschungsgemeinschaft (Germany) under grants TR-27 and EXC-153, as well as by the MSES of Croatia, is also acknowledged.

References

[1] R. D. Peccei and H. R. Quinn, *Constraints imposed by CP conservation in the presence of instantons*, *Phys. Rev.* **D16** (1977) 1791–1797.

[2] R. D. Peccei and H. R. Quinn, *CP conservation in the Presence of Instantons*, *Phys. Rev. Lett.* **38** (1977) 1440–1443.

[3] S. Weinberg, *A New Light Boson?*, *Phys. Rev. Lett.* **40** (1978) 223–226.

[4] F. Wilczek, *Problem of Strong P and T Invariance in the Presence of Instantons*, *Phys. Rev. Lett.* **40** (1978) 279–282.

[5] P. Sikivie, *Axion cosmology*, *Lect.Notes Phys.* **741** (2008) 19–50, [astro-ph/0610440].

[6] O. Wantz and E. P. S. Shellard, *Axion Cosmology Revisited*, *Phys. Rev.* **D82** (2010) 123508, [arXiv:0910.1066].

[7] P. Sikivie, *Experimental tests of the invisible axion*, *Phys. Rev. Lett.* **51** (1983) 1415.

[8] G. Raffelt and L. Stodolsky, *Mixing of the Photon with Low Mass Particles*, *Phys.Rev.* **D37** (1988) 1237.

[9] D. M. Lazarus *et. al.*, *A Search for solar axions*, *Phys. Rev. Lett.* **69** (1992) 2333–2336.

[10] S. Moriyama *et. al.*, *Direct search for solar axions by using strong magnetic field and X-ray detectors*, *Phys. Lett.* **B434** (1998) 147, [hep-ex/9805026].

[11] K. Zioutas *et. al.*, *A decommissioned LHC model magnet as an axion telescope*, *Nucl. Instrum. Meth.* **A425** (1999) 480–489, [astro-ph/9801176].

[12] I. G. Irastorza, F. Avignone, S. Caspi, J. Carmona, T. Dafni, *et. al.*, *Towards a new generation axion helioscope*, *JCAP* **1106** (2011) 013, [arXiv:1103.5334].

[13] J. Isern, E. Garcia-Berro, L. Althaus, and A. Corsico, *Axions and the pulsation periods of variable white dwarfs revisited*, *Astron. Astrophys.* **512** (2010), no. A86 86, [arXiv:1001.5248].

[14] J. Jaeckel and A. Ringwald, *The Low-Energy Frontier of Particle Physics*, *Annual Review of Nuclear and Particle Science* **60** (Nov., 2010) 405–437.

[15] A. Ringwald, *Exploring the Role of Axions and Other WISPs in the Dark Universe*, *Phys.Dark Univ.* **1** (2012) 116–135, [arXiv:1210.5081].

[16] O. K. Baker, M. Betz, F. Caspers, J. Jaeckel, A. Lindner, A. Ringwald, Y. Semertzidis, P. Sikivie, and K. Zioutas, *Prospects for searching axionlike particle dark matter with dipole, toroidal, and wiggler magnets*, *Phys. Rev. D* **85** (Feb, 2012) 035018.

[17] I. G. Irastorza and J. A. Garcia, *Direct detection of dark matter axions with directional sensitivity*, *JCAP* **1210** (2012) 022, [arXiv:1207.6129].

[18] D. Horns, J. Jaeckel, A. Lindner, A. Lobanov, J. Redondo, *et. al.*, *Searching for WISPy Cold Dark Matter with a Dish Antenna*, *JCAP* **1304** (2013) 016, [arXiv:1212.2970].

[19] I. Shilon, A. Dudarev, H. Silva, and H. Kate, *Conceptual Design of a New Large Superconducting Toroid for IAXO, the New International AXion Observatory*, *IEEE Trans. Appl. Supercond.* **23** (2012) [arXiv:1212.4633].

[20] J. E. Koglin *et. al.*, *Hard x-ray optics: from HEFT to NuSTAR*, *Proc SPIE* **856-867** (2004).

[21] A. Tomas, S. Aune, T. Dafni, G. Fanourakis, E. Ferrer-Ribas, *et. al.*, *CAST microbulk micromegas in the Canfranc Underground Laboratory*, *Phys.Procedia* **37** (2012) 478–482, [arXiv:1208.5690].

[22] J. García *et. al.*, *'Low background x-ray detection with Micromegas for axion research', talk at this conference*, .

Journal of Physics: Conference Series **460** (2013) 012003 doi:10.1088/1742-6596/460/1/012003

Low-background X-ray detection with Micromegas for axion research

J.A. García[1], T. Dafni[1], A. Diago[1], E. Ferrer-Ribas[2], J. Galán[2], A. Gardikiotis[3] I. Giomataris[2], J. G. Garza[1], F. J. Iguaz[1], I. G. Irastorza[1], I. Ortega[1], T. Papaevangelou[2], A. Rodríguez[1], A. Tomás[1], T. Vafeiadis[4], S. C. Yildiz[5]

[1]Laboratorio de Física Nuclear y Altas Energías, Universidad de Zaragoza, Zaragoza, Spain
[2]IRFU, Centre d'Études Nucléaires de Saclay (CEA-Saclay), Gif-sur-Yvette, France
[3]Physics Department, University of Patras, Patras, Greece
[4]European Organization for Nuclear Research (CERN), Genève, Switzerland
[5]Dogus University, Istanbul, Turkey

E-mail: jagarpas@unizar.es

Abstract. We have studied low-background techniques in order to improve the background level for Micromegas detectors. These detectors are good candidates for rare event searches thanks to the low mass and the radiopurity of the materials used in the construction; moreover they have good discrimination capabilities. The motivation of these studies is the reduction of the background level in the CAST experiment where three of the four detectors operating currently are Micromegas of the microbulk type. The last result of the experience acquired in low-background techniques has been the reduction of the background level by a factor 4.5 for two detectors in the CAST experiment, corresponding to an improvement of a factor 2 in signal strength.

1. Axion searches: CAST and IAXO

Axions could be produced in the Sun via the Primakoff effect and can be detected in the laboratory as x-rays via the inverse process in a strong magnetic field. This is the so-called helioscope technique [1], employed by the CAST experiment and which will also be exploited in the future IAXO.

1.1. The CAST experiment

The CERN Axion Solar Telescope experiment [2] is looking for Axion Like Particles (ALPs) since 2002. Using a LHC prototype dipole magnet which provides a magnetic field of 9T and is pointing at the Sun two times per day (during sunrise and sunset) around 1.5 hours each side. The signal of axions will be an excess of counts during the tracking in the X-ray detectors situated at the end of the four magnet bores. Three of the four detectors installed currently at CAST are Micromegas of the microbulk type [3].

CAST finished in 2011 the research program which includes a vacuum phase [4] for axion masses up to $m_a \leq 0.02$ eV and a buffer gas phase [5] [6] to increase the sensitivity to higher axion masses $0.02 \geq m_a \leq 1$eV which was divided in two parts respectively with ^{4}He and ^{3}He as buffer gas. Currently CAST is revisiting ^{4}He phase and will revisit vacuum phase obtaining a better sensitivity than in previous data taking due to a further reduction of the background levels in the detectors. The originalities of CAST are the use of an X-ray telescope to increase the signal to background ratio, and the use of low-background techniques which will be described in section 2.

1.2. The International AXion Observatory
IAXO, the International Axion Observatory [7] [8], will improve CAST's sensitivity by more than one order of magnitude. IAXO will exploit all the innovations of CAST with a larger toroidal magnet built for axion physics and all bores will be equipped with an X-ray telescope; moreover it will require low-background detectors. The Conceptual Design Report and a Letter of Intent for IAXO are under preparation.

2. Low-background techniques
CAST microbulk Micromegas exploit three different background strategies:

- *Low intrinsic radioactivity:* Microbulk Micromegas detectors are intrinsically radiopure due to the low mass and clean materials that are used in the construction, mainly copper, Plexiglas and kapton. The radiopurity of these detectors has been measured in the Canfranc Underground Laboratory (LSC) by a germanium detector [10] obtaining excellent results.
- *Signal topology:* The axion signature will be an excess of X-ray in the energy range from 1 to 10 keV. This kind of events in our detectors are point-like events and daily calibrations with ^{55}Fe source define the characteristic parameters of X-ray like events [9]. Micromegas detectors have two different readouts:
  - The pulse induced in the mesh, which gives time information and characteristic observables like *risetime, width, amplitude and integral.*
  - The charge collected in the strips of the anode, obtaining spatial resolution and defining *clusters* as a group of consecutive fired strips, allowing to define alternative observables in the strip plane like *cluster size, shape and charge.*

  Finally the combination of the different parameters of the observables during the calibrations define selection criteria for the background events obtaining excellent discrimination capabilities.
- *Shielding:* It is made of different layers of materials: external lead shielding and inner copper. The shielding at CAST is continuously evolving and recently has been improved for two detectors at CAST. The upgrade includes an active muon veto which will be described in section 3.1.

2.1. Measurements at the LSC
A specific set-up to evaluate the different background sources is a replica of a CAST microbulk detector installed at the Canfranc Underground Laboratory (LSC). The LSC is situated at Canfranc (Huesca) in the Spanish Pyrenees with a depth of 2500 m.w.e. where the muon flux is reduced by a factor 10^4 relative to surface. The set-up (Figure 1 left) is composed by a shielding of 10 cm of external lead and an inner copper layer with a thickness of 2.5 cm. Moreover nitrogen is flushed close to the detector to avoid Radon. In these conditions [11] [12] the background level is $< 2 \times 10^{-7}$ c keV^{-1}cm^{-2}s^{-1} in the $[2-7]$ keV range (Figure 1 right), 30 times lower than nominal CAST background $\sim 6 \times 10^{-6}$ c keV^{-1}cm^{-2}s^{-1} (before 2012 shielding upgrade).

Journal of Physics: Conference Series **460** (2013) 012003 doi:10.1088/1742-6596/460/1/012003

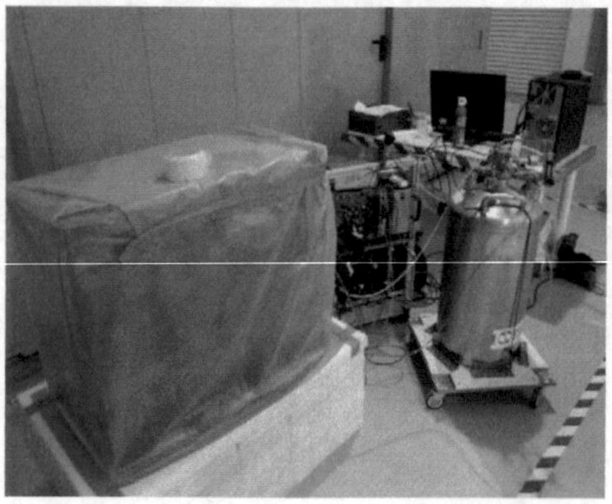

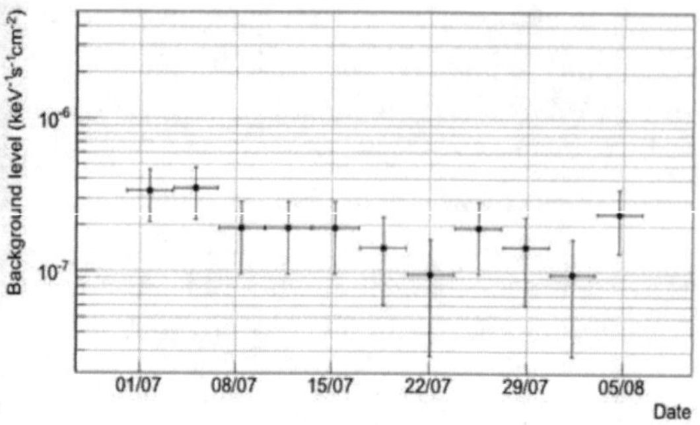

Figure 1. Left: Set-up with 10 cm lead shielding in the LSC. Right: background level evolution in the LSC

2.2. Simulations

The motivation of the simulations is to reproduce the background level observed in the Micromegas detectors at CAST, where the environmental $\gamma's$ flux has been experimentally measured and used as input for the simulations. The complete CAST set-up geometry has been implemented in GEANT4 to simulate physical events like ^{55}Fe calibration X-rays or environmental γ's. Afterwards with *RESTSoft*, a code developed at the University of Zaragoza, the drift and the diffusion in the chamber is simulated using Magboltz for the gas parameters. Finally the electronic response of the detector is implemented. A comparison between simulated and experimental data is presented in Figure 2.

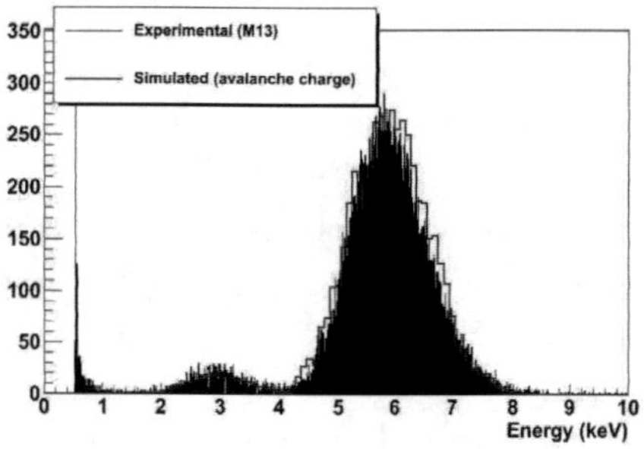

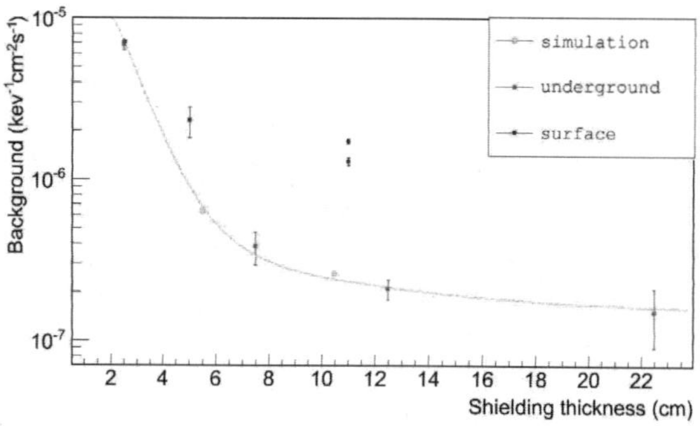

Figure 2. Left: Comparison between a ^{55}Fe calibration and a simulated one. Right: Background level in function of the shielding thickness for measurements at surface (blue dots), LSC (red dots) and simulations (green dots)

From the results shown on the right part of Figure 2 we conclude that the background level at surface has an important contribution of cosmic muons. This hypothesis has been confirmed by simulations, the background level versus the thickness of the shielding at the LSC (where muons are suppressed) can be reproduced by simulated data only using external γ's as input. But this behaviour cannot be reproduced by the measurements at surface level due the contribution of cosmic muons. These results motivated the installation of a cosmic veto at CAST.

3. Background levels at CAST

The Micromegas detectors at CAST have reduced the background level by a factor ~ 50 from the beginning of the experiment, with several upgrades which include the detector design from classic Micromegas to bulk and microbulk technology, the shielding thickness and the analysis of the data. Two remarkable improvements in the background reduction were related with a

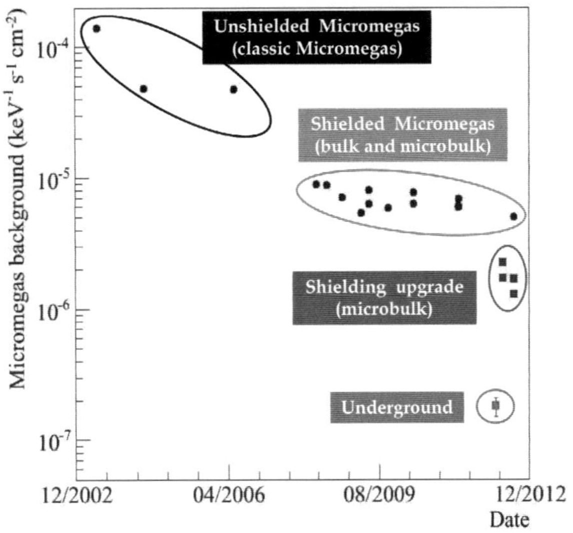

Figure 3. Evolution of the background level in Micromegas detectors at CAST (black and blue dots) and background level in special conditions at LSC (red dot)

shielding upgrade. In 2007 was the first installation of a shielding for the detectors, made by two layers of 25 mm of external lead shielding and 5 mm of inner copper improving the background 4.5 times (dots inside the green ellipse in Figure 3). In 2012 a shielding upgrade was made in the sunset detectors (blue dots in Figure 3) that will be detailed in next section.

3.1. Shielding upgrade of 2012

The shielding upgrade of 2012 for the Micromegas detectors at CAST in the sunset side is the result of the experience acquired in low-background techniques, the improvements includes:

- New shielding design made of 10 cm of external lead and 1 cm of inner copper. Now the shielding is more compact, improving the shielding around the pipes and magnet bores (Figure 4 left).
- The materials close to the detectors are radiopure (mainly copper and kapton) and have been carefully cleaned. Moreover all the steel pieces in the inner part of the shielding have been replaced by copper to prevent the $[5-7]$ keV fluorescence peaks.
- Active muon veto: As shown in section 2.2 muons have an important contribution at surface level and can induce fluorescence in the materials close to the detector. With this upgrade we obtained a 25% of background reduction in the $[2-7]$ keV energy range (Figure 4 right) with a non optimum veto (only 45% of coverage due to mechanical limitations), the veto coverage will we improved in next data taking campaigns.

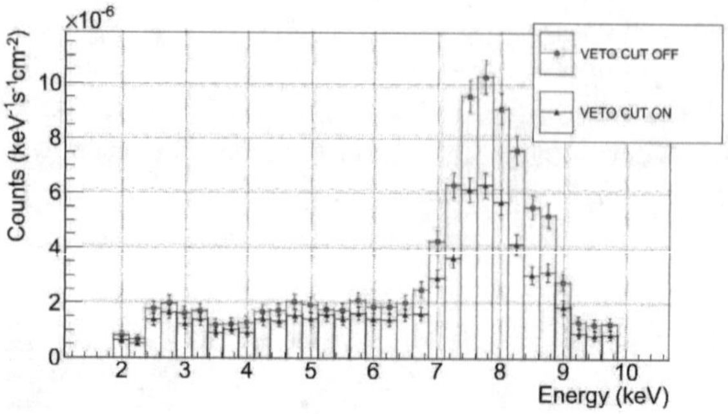

Figure 4. Left: Shielding upgrade for the Micromegas detectors of the sunset side in the CAST experiment. Right: Background spectra after the shielding upgrade before (red dots) and after the veto cut (blue triangles). It is remarkable the reduction of the fluorescence peak of the copper (8 keV) after the veto cut as a hint of the induced fluorescence by muons.

After the upgrade the background level reached was $\sim 1.5 \times 10^{-6}$c keV^{-1}cm^{-2}s^{-1}, a factor 4.5 of reduction in comparison with previous campaigns of data taking.

4. Future prospects

CAST will revisit the vacuum phase in 2013 and a new X-ray telescope will be installed in the sunrise Micromegas line, increasing the sensitivity of CAST due to the better signal-to-noise ratio. Also a new detector and shielding design has been proposed, motivated by the results obtained in the sunset upgrade and to improve the weak points of previous set-ups. In the new design the body and the support of the chamber is made completely of copper with a thickness of 2.5 cm also used as inner shielding; the window is smaller due the telescope and it will be tighter, the external shielding is made by 10 cm of lead and a cosmic scintillator will be installed. Another improvement is the new DAQ electronics, based on the AFTER chip [13] replacing the actual Gassiplex-based electronics [14] in all Micromegas detectors at CAST. In the AFTER-based electronics the strips pulses are digitized and the pulse shape analysis could be extended to every strip and a further background reduction could be archieved.

With the upgrades described previously CAST could improve the vacuum results down to a sensitivity of $g_{a\gamma} < 5.9 - 6.3 \times 10^{-11}$ GeV^{-1} (Figure 5) corresponding to a factor 4-5 in signal strength.

The long term prospects are focused on the International AXion Obervatory. An ultra-low background Micromegas is required for IAXO, with the goal of a background level of 10^{-7} c keV^{-1}cm^{-2}s^{-1} at surface level, down to 10^{-8} c keV^{-1}cm^{-2}s^{-1} if possible. With a dedicated magnet, optics and ultra-low background detectors, IAXO could improve the sensitivity of CAST in a factor $\sim$ 18 (10^5 in signal strength) [15]. Thanks to IAXO a big part of the QCD axion model could be explore next decade.

5. Conclusions and next steps

The work presented here is focused on reducing the background levels in Micromegas detectors and further developing low-background techniques. The goal is to build a background model for Micromegas detectors using all the information acquired with the measurements at the LSC and in the laboratory at surface level. The simulations are an important input to evaluate the different contributions in the final background of external γ's, muons and the intrinsic radioactivity of the detector. The proposed upgrade for CAST in the Sunrise line will be

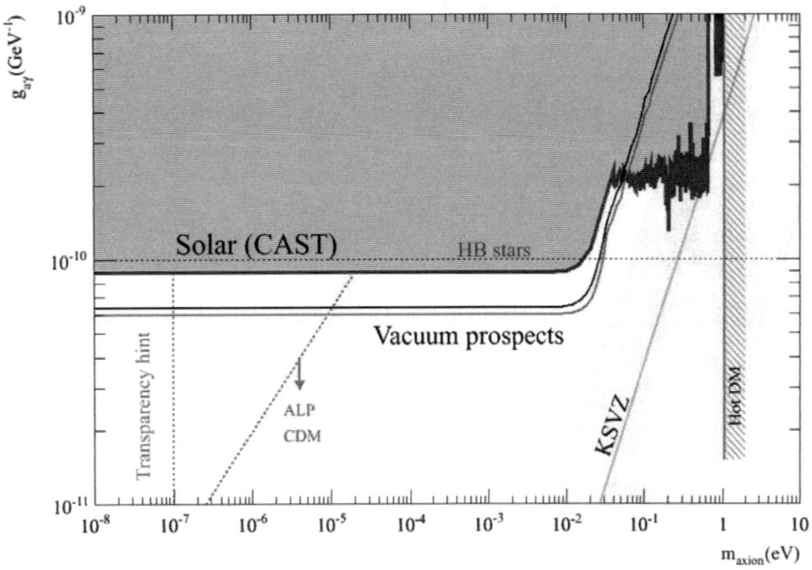

Figure 5. Expected sensitivity at CAST with the upgrades proposed for 9 months of data taking in vacuum. Two assumptions in the background for the Micromegas detectors have been computed a conservative scenario with a background level 1.5 $\times$ 10^{-6} c keV^{-1}cm^{-2}s^{-1} (black line) corresponding to a sensitivity of $g_{a\gamma}$ < 6.34 $\times$ 10^{-11} GeV^{-1} and a optimistic scenario 8 $\times$ 10^{-7} c keV^{-1}cm^{-2}s^{-1} (red line) with a sensitivity of $g_{a\gamma} < 5.94 \times 10^{-11}$ GeV^{-1}. Plot taken from [16].

important for the development of low-background techniques and it will be crucial in order to improve the background levels required for IAXO.

Acknowledgments

We thank CAST Collaboration for the contribution to this paper and their effort in the daily work at CERN. We acknowledge support from the Spanish Ministry of Science and Innovation (MICINN) under contract ref. FPA2011-03456, as well as the support from the European Commission under the European Research Council T-REX Starting Grant ref. ERC-2009-StG-240054 of the IDEAS program of the 7th EU Framework Program.

References

[1] Sikivie P 1983 *Phys. Rev. Lett.* **51** 1415
[2] Zioutas K et al 1999 *Nucl. Instrum. and Meth. A* **425** 27
[3] Papaevangelou T et al 2010 *Proc. of 'The* 1st *International Conference on Micro Patern Gaseous Detectors'* (Crete)
[4] Zioutas K et al 2005 *Phys. Rev. Lett.* **94** 121301
[5] Arik E et al 2009, *JCAP* **0902** 008
[6] Aune S et al 2011 *Phys. Rev. Lett.* **107** 261302
[7] Irastorza I G et al 2011 *Proc. of 'The* 7th *Patras workshop on axions, WIMPs and WISPs'*(Mykonos)
[8] Irastorza I G et al 2011 *JCAP* **1106** 013
[9] Tomas A et al 2011 *X-Ray Spectrom.* **40** 240-246
[10] Cebrian S et al 2011 *Astropart. Phys.* **34** 354
[11] Tomás A et al 2012 *Physics Procedia* **37** 478
[12] Galán J A et al 2012 *EAS Publications Series* **53** 155

[13] Baron P et al 2008 *IEEE Trans. Nucl. Sci* vol 55 no 3 pp 1744-1752
[14] Santiard J C and Marent K 1999 5th *Conference on Electronics for LHC Experiments* (Snowmass) pp 431-435
[15] Irastorza I G et al 2012 *Proc. of 'The* 6th *Symposium on large TPCs for low energy rare event detection'* (Paris)
[16] Papaevangelou 2012 *Status report of the CAST Experiment and Running in 2013-2014* CERN-SPSC-2012-028

Journal of Physics: Conference Series **460** (2013) 012004 doi:10.1088/1742-6596/460/1/012004

GridPix as a candidate for the future of CAST

J Kaminski, C Krieger, Y Bilevych, T Krautscheid and K Desch
Physikalisches Institut, Universität Bonn, Nussallee 12, 53115 Bonn, Germany
E-mail: kaminski@physik.uni-bonn.de

Abstract. A central part of the CAST upgrade program is the further reduction of the background rates and the extension of the experiment's sensitivity to lower energies. Both aims could be reached with a highly pixelized readout of the detector as, for example, with an InGrid, the combination of a Timepix ASIC and a Micromegas grid. The high granularity and efficiency of such a device make it possible to detect and to distinguish single primary electrons. Therefore, we have studied the energy resolution by electron counting and the separation of photon events from tracks with an event shape analysis.

1. Introduction

The CAST experiment [1] searches for axions and other axion-like particles emitted by the sun. The axions are supposed to convert in the strong magnetic field of CAST into X-ray photons and can then be recorded by X-ray detectors. To improve the current limits and to also allow the search for other particles such as chameleons [2], both the background rates must be further reduced (see also [3]) and the threshold of the energy spectrum must be lowered. The background signals consist mostly of tracks from radioactive decays and from cosmic rays. To suppress these events, detector materials with low intrinsic radioactivity have to be chosen and the detectors have to be well shielded. Besides, a reliable separation of photon and background signals has to be ensured in the off-line analysis on an event-by-event basis. If the events are recorded with a very high granularity, an event shape analysis is a promising approach to distinguish signals from background. We have, therefore, constructed a test detector read out with an InGrid [4], which is a combination of a Timepix ASIC [5] and a Micromegas gas amplification stage [6]. Measurements both with a ^{55}Fe-source and without have demonstrated a high energy resolution and a powerful background suppression. Because of the limited geometrical size, this detector is a good candidate for an X-ray detector behind an X-ray telescope at CAST.

2. InGrid detector

The cylindrical detector has a drift length of 2 cm and an inner diameter of 8 cm. Most of the components are made of aluminum and HV insulation has been done with Kapton®. The cathode features a 1 mm^2 entrance window for X-rays, which is covered with a 50 µm thin Kapton window. The anode has an opening where the InGrid is mounted (see figure 1). This device uses the Timpix ASIC both as a readout plane, where the bump bond pads of the chip are used as charge collection pads, and as a digitization chip. For protection of the chip's circuitry the readout area is covered with a resistive layer of Si_XN_Y [7]. Besides, a Micromegas mesh is mounted on the chip with post-processing techniques, which ensure a very precise fabrication and alignment of the grid holes with the pads of the Timepix ASIC. Thus, both the grid holes

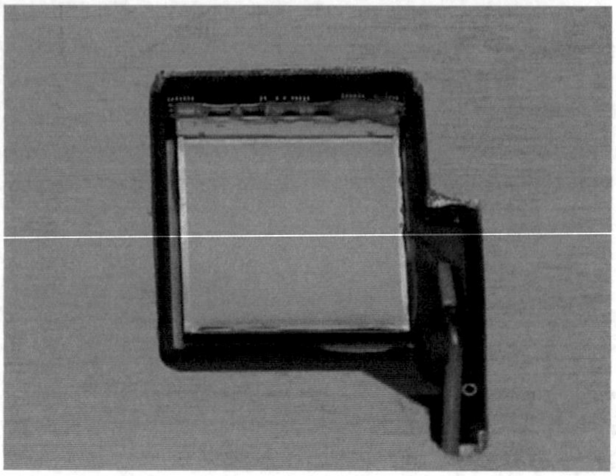

Figure 1. InGrid mounted in the anode of the detector.

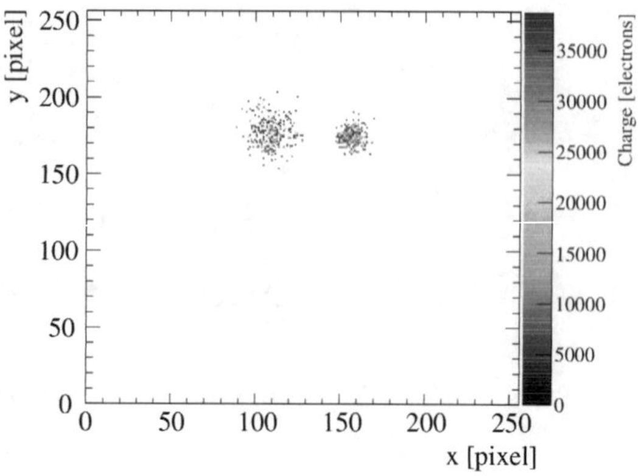

Figure 2. Conversion of two X-ray photons at different drift distances.

and the pads have a pitch of 55 × 55 μm^2 and cover an area of about 2 cm^2. The Timepix ASIC was controlled and read out by the MUROS 2.1 [8] and the Pixelman software [9].

During data taking, the detector was flushed with a gas mixture of argon:isobutane (95:5) and the gas pressure was kept at a few mbar above atmospheric pressure. An electric field of 225 V/cm was applied between the cathode and the anode and the voltage on the grid was varied from 290 V up to 400 V.

3. Detector performance

For testing the detector's performance, a ^{55}Fe source was placed on the detector and a chromium foil was used to suppress the 6.1 keV line of the source. Figure 2 shows an example event, where two X-ray photons converted at different drift distances. The energy of the converted photon can be extracted by either summing up the charge of all pixels associated with one photon or by simply counting the number of activated pixels, which corresponds to counting the number of primary electrons. If the diffusion is high enough for spreading the charge sufficiently, the second approach gives a better energy resolution. For a grid voltage of 350 V an energy resolution of $\sigma_E/E = (5.23 \pm 0.03)$ % was measured.

In Figure 3 a background track is shown passing through the detector. The track width reveals that the lower left end passed through the cathode and the upper right end passed through the InGrid. To distinguish tracks and X-ray conversions all events are reconstructed twice: Once with the hypothesis of being a photon event, and the second time with the hypothesis of being a track. In both cases characteristic geometric parameters are calculated. For example, in the track analysis the number of hits per track length, kurtosis and eccentricity along the long axis and the track length are determined. The distribution of these parameters have also been calculated for a pure data sample of photon events and background events (see track length distribution in figure 4). During the analysis the value for each event is compared to the two sample distributions and a likelihood ratio is used for the final decision to accept the event as a photon or to discard it as a track. The resulting discrimination power depends of course on the cut value of the likelihood parameter, but for an efficiency of (95.3 ± 0.3) % for 5.9 keV X-ray photons and a track mis-acceptance rate of (1.1 ± 0.6) % the background could be suppressed by a factor of about 120. More details of the analysis are described in reference [10].

Journal of Physics: Conference Series **460** (2013) 012004 doi:10.1088/1742-6596/460/1/012004

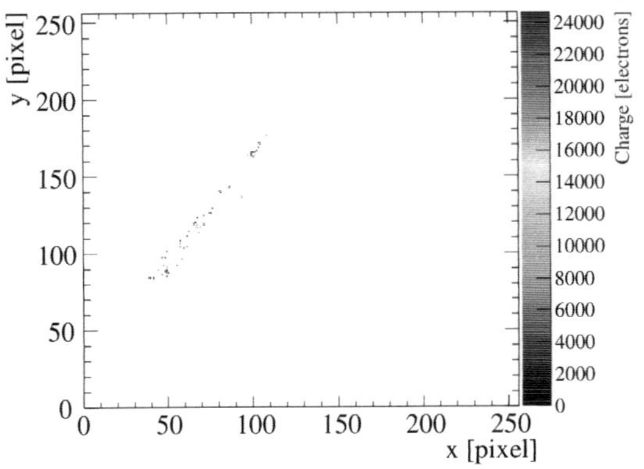

Figure 3. Track of a cosmics ray traversing the complete drift length of the detector.

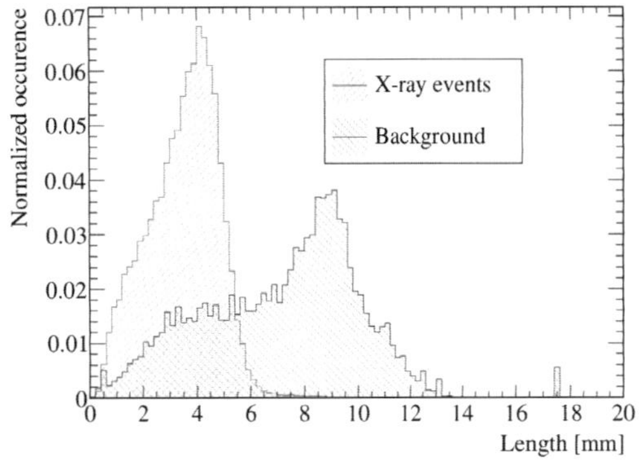

Figure 4. Distribution of the track length for both an X-ray photon data sample and a background data sample.

4. Further plans

Several detector components have been rebuilt from more radiopure materials like acrylic glass. Currently, studies on entrance windows with very low material budget are ongoing and in the close future also the signal shall be decoupled from the grid. The time structure of this signal will be used as an additional parameter in the likelihood to improve the separation power for tracks that enter the detector in a steep angle with respect to the read out.

Since the production of the InGrids at the University of Twente was limited to a small number per production cycle, a new process has been established at the Fraunhofer Institute IZM at Berlin, where a wafer-based production yields up to 100 InGrids per production cycle. The third batch has been delivered in the fall of 2012 and first tests confirm that the grids are of very high quality. A picture taken with an optical microscope is shown in figure 5. Tests with an ^{55}Fe source demonstrate a very good behavior for a wide range of gas amplification (grid voltages of 290 V to 400 V). An energy spectrum for a grid voltage of 350 V is shown in figure 6, which gives an energy resolution of $\sigma_E/E = (5.6 \pm 0.1)$ %. For higher voltage settings thousands of discharges were observed, but most chips were not affected.

Figure 5. Close up view of InGrid.

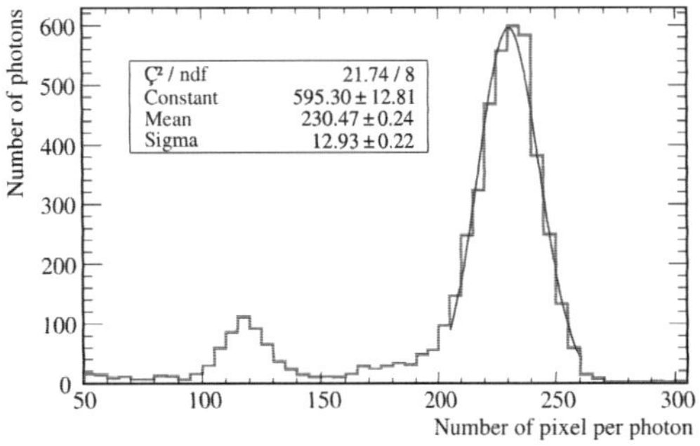

Figure 6. Energy spectrum of an ^{55}Fe source.

Finally, also a new readout system for the Timepix ASIC is designed at the University of Bonn. It is based on the Scalable Readout System [11] of the RD51-collaboration and because of its open source approach will facilitate the implementation of additional features such as the aforementioned grid signal processing.

5. Acknowledgments

We would like to thank our colleagues Y. Giomataris, E. Ferrer Ribas, T. Papaevangelou, H. van der Graaf, and J. Schmitz for their advice in developing this detector and commissioning it.

References

[1] Zioutas K *et al.* 1999 A decommissioned LHC model magnet as an axion telescope *Nucl. Instr. and Meth.* A **425** 480-7

[2] Brax P, Lindner A, Zioutas K 2011 Chameleon foreCAST *Preprint* arXiv:1110.2583v3 [hep-ph]

[3] Garcia J 2012 Low background X-ray detection with Micromegas for axion research *J. Phys.: Conf. Series* of this conference

[4] Chefdeville M *et al.* 2006 An electron-multiplying 'Micromegas' grid made in silicon wafer post-processing technology *Nucl. Instr. and Meth.* A **556** 490-4

[5] Llopart X, Ballabriga R, Campbell M, Tlustos L and Wong W 2007 Timepix, a 65k programmable pixel readout chip for arrival time, energy and/or photon counting measurements *Nucl. Instr. and Meth.* A **581** 485-94

[6] Giomataris Y, Rebourgeard P, Robert J P and Charpak G 1996 Micromegas: a high-granularity position-sensitive gaseous detector for high particle-flux environments *Nucl. Instr. and Meth.* A **376** 29-35

[7] Bilevych Y *et al.* 2011 Spark protection layers for CMOS pixel anode chips in MPGDs *Nucl. Instr. and Meth.* A **629** 66-73

[8] San Segundo Bello D, van Beuzekom M, Jansweijer P, Verkooijen H and Visschers J 2003 An interface board for the control and data acquisition of the Medipix2 chip *Nucl. Instr. and Meth.* A **509** 164-70

[9] Turecek D, Holy T, Jakubek J, Pospisil S and Vykydal Z 2011 Pixelman: a multi-platform data acquisition and processing software package for Medipix2, Timepix and Medipix3 detectors *JINST* **6** C01046

[10] Krieger C, Kaminski J and Desch K InGrid-based X-ray detector for low background searches submitted to *Nucl. Instr. and Meth.*

[11] Martoiu S, Muller H and Toledo J 2011 Front-end electronics for the Scalable Readout System of RD51, *IEEE NSS Conf. Rec. 2011* 2036-8

Journal of Physics: Conference Series **460** (2013) 012005 doi:10.1088/1742-6596/460/1/012005

The LUX Dark Matter Search – Status Update

S. Fiorucci, for the LUX Collaboration

Brown University, Dept. Of Physics, 182 Hope St, Providence RI 02912

fiorucci@brown.edu

Abstract. We report on the design, construction and commissioning of the Large Underground Xenon (LUX) dark matter detector at the Sanford Laboratory in Lead, SD, USA. From its inception in 2007, to its construction at a surface laboratory in lead in 2009-2010, its operation in 2011, and its re-installation 1 mile underground in 2012, we review the relevant achievements already obtained and give an outlook on how LUX will become the most sensitive detector in the field in 2013.

1. The LUX Detector

The large Underground Xenon (LUX) detector is a dual-phase xenon time projection chamber for the direct detection of dark matter particle interactions. With a total target mass of 370 kg, LUX is currently the largest detector of its kind. Aside from its larger size, the LUX detector possesses a number of particularities which contribute to make it extremely competitive. For more details on all technical aspects of the LUX detector the reader is encouraged to consult the recently published overview paper in Ref. [1]. Of particular note are the following:

- A double-walled cryostat made of very-low radioactivity titanium, the product of a LUX research effort [2], contains the xenon and internals.
- 122 Photo-Multiplier Tubes (PMT) with a 2-inch circular window are used in two arrays on top and bottom of the active xenon volume, in order to collect light signals. Their low radioactivity content [3] and excellent quantum efficiency allow LUX to control backgrounds and obtain outstanding light collection efficiency, measured at > 2× better than the nearest competitor –see section 3 below for details on results.
- A cryogenic system based on the liquid nitrogen thermosyphon technology, provides several hundred watt of cooling power at liquid xenon temperatures [4]. Dual phase heat exchangers lower the need for liquid nitrogen dramatically, with an efficiency measured at > 94% for a flow rate up to 350 kg/d.
- An automatic, in-line xenon sampling system allows the collaboration to verify the purity of the xenon at various points in the circulation loop, with sensitivities to levels in N_2, O_2, CH_4 and Kr down to 1 ppb, 160 ppt, 60 ppt and 0.5 ppt, respectively [5,6].
- A mechanism is in place for doping the xenon on demand within the circulation loop, with radioactive isotopes such as ^{83m}Kr or ^{3}H [7]. It provides a controlled source of low-energy interactions distributed throughout the entire xenon volume, and is a very effective tool to calibrate the detector.
- Shielding against ambient gammas and neutrons is provided by an 8 m diameter, 6 m tall tank filled with constantly purified and de-radonized water. The water tank is also

instrumented as a Cherenkov veto, allowing the rejection of events in coincidence with rare cosmic muons traversing the tank.

2. Commissioning of the LUX detector

The LUX collaboration was formed in 2007 and fully funded jointly by NSF and DOE in 2008. Its history is intimately linked to that of the new Sanford Underground Research Facility (SURF) in Lead, South Dakota.

For the first three years, the lab did not exist as a research facility and the R&D effort for LUX was focused at Case Western Reserve University with the "LUX 0.1" program [8]. This allowed us to build and operate a full-size prototype of the LUX detector, with an emphasis on cryogenics, gas circulation, light readout and slow-control. It also provided a tremendous learning and training opportunity for the students and postdocs who would eventually build and operate the real detector.

By June 2010, a dedicated Surface Laboratory at SURF was completed to host the construction and commissioning of the LUX detector. This facility is as much as possible a 1:1 replica of the underground laboratory, including a class 1,000 clean room and a (smaller) water tank. It allowed the collaboration to fully test most systems while the underground facility was being excavated and outfitted. The LUX detector was assembled from the inside out between 2010 and 2011 within the clean room, while external systems, such as the xenon circulation loop, data acquisition electronics, xenon storage system, and nitrogen distribution, were built and tested just outside. In May 2011 the detector assembly was completed with 30 PMTs installed. A first cryogenic test was performed, successfully bringing the internal temperature down to 190 K and keeping it stable over a week, with the detector filled with argon gas. By October 2011 the LUX detector had been outfitted with the rest of the 122 PMTs and final upgrades to internal and external systems were completed. Between October 2011 and February 2012 LUX undertook "Run 2", during which the detector was immersed in the water shield, the full 370 kg xenon load was condensed, circulation/purification was started and data were acquired. The results were very satisfying (see section 3 below and Ref. [9]). Only a very limited number of fixes and upgrades to perform before underground re-deployment, were identified. These were corrected between March and July 2012.

In June and July 2012, the entire experiment was packed up and prepared for transport underground. The Davis Laboratory, 4850 feet underground, was completed in June 2012. The two-story laboratory space includes the 8 m diameter, 6 m tall water tank, meant to provide shielding for the LUX detector, as well as a clean room, a control room, a liquid nitrogen storage and a water filtration unit. Between July and September 2012 all of LUX was transported and reinstalled in the Davis Laboratory. The water tank was filled with water in late October, and LUX started taking Cherenkov data with the detector at vacuum. On December 10, LUX officially started "Run 3" by introducing ultra-pure xenon gas into the detector, with the goal to cool down and condense the full 370 kg by the end of January 2013.

3. Performance

LUX "Run 2" at the SURF Surface Laboratory achieved a number of remarkable results and demonstrated that the experiment was ready to move underground and start looking for dark matter. For more details the reader is encouraged to consult Ref. [9], but the following points are worth emphasizing:

- LUX achieved stable operation for over 100 days of running, circulating xenon at 35 slpm (equivalent to ~300 kg/day) with >98% heat exchanger efficiency, resulting in a total heat load <5 W.
- After 2 months of circulating, LUX achieved a xenon purity of 205 μs, corresponding to an electron drift length of ~25 cm, or half a detector. While not optimal, this number comes with the major caveat that a leak in the circulation path was discovered at the beginning of the run, leading to non-uniform mixing of the xenon and relatively ineffective purification.

- Light collection was measured with no applied electric field, at 662 keV (^{137}Cs gamma line), at 8.0 photo-electron / keV. This number is more than a factor 2 higher than that reported under similar conditions by competing experiments [10]. In terms of absolute probability for detection of scintillation photons, this corresponds to α_1 ~15% [11]. It is very encouraging both for obtaining a low energy threshold in underground running, and for the reflectivity of PTFE in liquid xenon, which has to be >97% in order to fit the data.
- LUX measured the positively-charged muon lifetime from tagged muon interactions in the detector at 2.16 ± 0.09 μs, in good agreement with the world average of 2.197 μs.
- The position of individual events was reconstructed with a resolution better than 5 mm in X and Y, based on an analysis of alphas emitted from Rn daughters plated out on the gate grid wires, and from a coincidence analysis within the whole xenon volume of ^{214}Bi – ^{214}Po radioactive decays.

4. Outlook

The LUX underground science program will consist of two steps. In the first step, after ~1 month of circulating xenon to improve the purity, we will run some initial calibrations and get an idea of the background levels. We will then run in dark matter search mode for a period of ~60 days. The exact duration will be decided in a large part by the observed background levels, and tuning of the acquisition parameters. This will be a non-blind search, and we expect to see very few electronic recoil events in the energy region of interest (in a 100 kg fiducial volume, before cuts), and no nuclear recoil events by a comfortable margin. The result of this short campaign should already make LUX the most sensitive dark matter detector in the world, and will be available in 2013.

In a second step, after a new round of extensive gamma and neutron calibrations, LUX expects to start a new 300 day dark matter search campaign. This new search will be blind and will rely on early data and calibration data to define the cuts and search region. The estimated sensitivity at the end of the experiment is a WIMP-nucleus cross-section smaller than 2×10^{-46} cm^2 for a WIMP mass of 40 GeV/c^2. More details on this estimate and what assumptions go into it can be found at the end of Ref. [9].

Acknowledgements

This work was partially supported by the U.S. Department of Energy (DOE) under award numbers DE-FG02-08ER41549, DE-FG02-91ER40688, DOE, DE-FG02-95ER40917, DE-FG02-91ER40674, DE-FG02-11ER41738, DE-SC0006605, DE-AC02- 05CH11231, DE-AC52-07NA27344, the U.S. National Science Foundation under award numbers PHYS-0750671, PHY-0801536, PHY-1004661, PHY-1102470, PHY-1003660, the Research Corporation grant RA0350, the Center for Ultra-low Background Experiments in the Dakotas (CUBED), and the South Dakota School of Mines and Technology (SDSMT). LIP-Coimbra acknowledges funding from Fundação para a Ciência e Tecnologia (FCT) through the project-grant CERN/FP/123610/2011. We gratefully acknowledge the logistical and technical support and the access to laboratory infrastructure provided to us by the Sanford Underground Research Facility (SURF) and its personnel at Lead, South Dakota.

References

[1] D.S. Akerib, et al., "The large Underground Xenon (LUX) Experiment," Nucl. Instrum. Meth. A 704 111-126 (2013), arXiv:1211.3788
[2] D.S. Akerib, et al., "Radio-assay of Titanium samples for the LUX Experiment," Submitted to Nucl. Instrum. Meth. A., arXiv:1112.1376
[3] D.S. Akerib, et al., "An Ultra-Low Background PMT for Liquid Xenon detectors," Nucl. Instrum. Meth A 10.1016 / j.nima.2012.11.020, arXiv:1205.2272
[4] A. Bolozdynya, et al., "Cryogenics for the LUX detector", IEEE Trans. Nucl. Sci. (2008)
[5] D.S. Leonard, et al., "A simple high-sensitivity technique for purity analysis of xenon gas," Nucl. Instrum. Meth. A 621 678-684 (2010). arXiv:1002.2742

[6] A. Dobi, et al., "Detection of krypton in xenon dfor dark matter applications," Nucl. Instrum. Meth. A 665 1-6 (2011). arXiv:1103.2714
[7] L.W. Kastens, et al., "A ^{83m}Kr Source for Use on Low-background Liquid Xenon Time Projection Chambers," Phys. Rev. C 80 045809 (2009). arXiv:0905.1706
[8] D.S. Akerib, et al., "The LUX Prototype Detector: heat Exchanger Development", accepted for publication in Nucl. Instrum. Meth. A (2013). arXiv:1207.3665
[9] D.S. Akerib, et al., "Technical Results from the Surface Run of the LUX Dark Matter Experiment," submitted to Astropart. Phys., arXiv:1210.4569
[10] E. Aprile, et al., "The XENON100 Dark Matter Experiment," Astropart. Phys. 35 573-590 (2012)
[11] P. Sorensen and C.E. Dahl, "Nuclear recoil energy scale in liquid xenon with application to teh direct detection of dark matter," Phys. Rev. D 83 063501 (2011). arXiv:1101.6080

Journal of Physics: Conference Series **460** (2013) 012006 doi:10.1088/1742-6596/460/1/012006

Columnar recombination: a tool for nuclear recoil directional sensitivity in a xenon-based direct detection WIMP search

D R Nygren

MS 50R5008 Lawrence Berkeley National Laboratory
1 Cyclotron Road, Berkeley CA 94705, USA

drnygren@lbl.gov

Abstract. A robust signal for sidereal anisotropy in nuclear recoils would support, perhaps more decisively than any other evidence, a discovery claim for a WIMP component of Dark Matter. I present a concept based on *columnar recombination in dense xenon gas*, sensing nuclear recoil direction relative to a TPC drift field. The central advance is that nuclear recoil directionality information is obtained through a comparison, event-by-event, of the ionization signal and recombination signal that are produced *prior* to drifting the track ionization. The optimum xenon density for this concept may be near ten bars, unlike conventional techniques that employ track visualization – with severe restrictions on gas density to about 1/10 bar. No restriction is imposed by diffusion during drift, facilitating the realization of a large monolithic room temperature xenon gas Time Projection Chamber at the ton-scale, with unprecedented sensitivity for both directionality and cross-section. Remarkably, the desired operating conditions for 0-νββ ^{136}Xe experiment may be identical.

1. Introduction

An attractive candidate for Dark Matter (DM) is the Weakly Interacting Massive Particle (WIMP). As a remnant of Big Bang era, WIMPs should have interaction strength characteristic of weak interactions, although WIMP mass and interaction strength are not strongly constrained by theory. WIMP masses might lie in a wide range from well above normal heavy nuclei down to perhaps 10 GeV/c^2. WIMP DM gravitationally captured in our galaxy is regarded as unlikely to co-rotate with the galactic plane. Independent of mass, WIMPs would have a truncated virial mean velocity of approximately 230 km/s. Earth's position in the galactic arm co-adds a velocity of ~220 km/s to a WIMP-nucleus collision in a terrestrial detector. A substantial anisotropy would thereby be induced in the distribution of nuclear recoil directions – most pronounced for the most energetic collisions. Local variations in the WIMP flow may be quite different from the simplest scenario, including even a negligible flux at earth, but we can only hope that Nature has not been that unkind to us. A recent review of the prospects for detecting directionality in WIMP flows can be found in [1].

The angle that the earth's rotation axis makes with the velocity of the galactic plane allows for automatic scanning with regard to putative WIMP flow direction from Cygnus. In figure 1, adapted from [1], the basic concept is illustrated.

Journal of Physics: Conference Series **460** (2013) 012006 doi:10.1088/1742-6596/460/1/012006

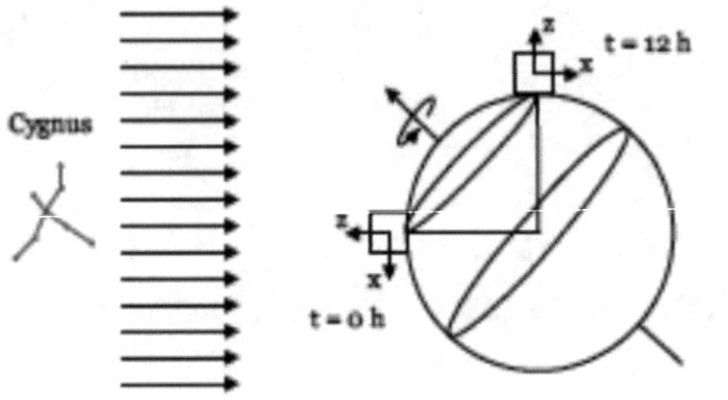

Figure 1. Under the assumption of a WIMP wind favoring the direction from Cygnus, a detector on earth observes sidereal anisotropy. At any latitude, the detector can be oriented to provide equal sensitivity. (figure is adapted from [1])

The discovery of sidereal anisotropy in nuclear recoils would provide strong and arguably compelling evidence for the WIMP nature of DM. At present, however, there is no compelling basis that DM is predominantly WIMPs or that WIMPs must have any of these properties. As yet, there exists no widely accepted evidence of their presence at earth.

In the standard WIMP scenario, the expected nuclear recoil energy spectra fall rapidly with energy. Such recoils, typically not more than a few tens of keV, fall well within the range of abundant and nasty backgrounds due to radioactivity and cosmic radiation. Despite these challenges, experimental limits on interaction cross-section versus WIMP mass have been pushed steadily and impressively lower.

In the event that evidence for nuclear recoils is obtained that cannot be explained with known processes, the question "What quality of evidence is required for acceptance that discovery has been made of the nature of approximately six times more matter than the visible universe?" will become central, but will likely remain without consensus. Then the search for directionality in such nuclear recoils will become of immediate and paramount interest. No known terrestrial background can mimic a directionality signal. It seems prudent to prepare for this eventuality.

To date, almost all approaches pursued for directionality sensitivity attempt to obtain a useful measure of nuclear recoil trajectory *after* drifting an ionization image of the nuclear recoil to a measurement plane with avalanche gain [1]. The degrading effects of diffusion during drift, avalanche gain, and visualization/ reconstruction noise can only be overcome by increasing the range of nuclear recoils through the use of a dilute gas target and the imposition of a maximum allowable drift length to constrain diffusion. These requirements limit the volume to ~1 m^3 or less, and active mass per detector to ~100 grams or less. In contrast, the absence of signals in non-directional direct detection detectors now in operation indicate that hundreds of kg – most popularly liquid xenon (LXe) or liquid argon (LAr) – may be required for adequate sensitivity, and tons are contemplated or near realization for the very near future. Between detectors without directional and those with directional sensitivity, a difference of at least three orders of magnitude in active mass exists; how can this gap be confronted?

A scheme that combines a large, monolithic, active mass with sensitivity to WIMP wind directionality would hence be most welcome. None have been previously conceived, discussed, or proposed, to the best of my knowledge. Here I ask: Can *columnar recombination* be "optimized" to sense nuclear recoil directionality efficiently in a large, massive, monolithic detector? I propose a scheme in which all information is collected in the form of optical signals, based on a high-pressure xenon gas electroluminescent (EL) TPC with a special (and possibly unique) molecular additive.

2. Columnar recombination as a tool to sense nuclear recoil direction

Recombination is the capture by an ion of an electron freed by an energetic ionizing particle, an ordinary process in ionized media and generally regarded as either a nuisance or of no special interest. Electron capture is likely when an electron wanders sufficiently near an ion; the relevant proximity for

capture is the Onsager radius, defined as $r_O \equiv e^2/\varepsilon E$. r_O is that distance between a positive ion and a free electron for which the potential energy (negative) is balanced by the electron's kinetic energy E. ε, the dielectric constant of the medium, incorporates polarization effects. The Onsager radius is well defined for any sub-excitation energy. To set the scale for a thermalized electron with thermal kinetic energy E = kT in xenon gas, $r_O \approx 70$ nm. In liquid xenon (LXe), polarization decreases r_O to 54 nm.
Jaffe published the first study of columnar recombination (CR) in 1913 [2]. The fundamental insight here is that, under the right conditions, sensitivity exists in recombination processes to the angle between a column of ionization and an electric field – supplied here by a TPC. Perhaps, this effect can be harnessed to substantiate the "flow" of WIMPs, should evidence build that Nature has chosen this model for dark matter.
A large angle between track and field leads electrons transversely away from the ion column. The electrons have relatively few encounters with ions, and a recombination signal is small relative to the ionization signal. Conversely, a small angle implies a higher level of recombination as the electrons drift more or less parallel to the ions, encountering many; a recombination signal is relatively large in comparison to the surviving ionization signal. The measure of directionality is found through a comparison, for each event, of a recombination signal "**R**" to the surviving ionization signal "**I**". The ratio **R/I** for each event carries the message of directionality.
The **R** signal is UV scintillation; the challenge is to maximize the detection efficiency and statistical precision of the **R** signal in a detector of interesting scale. As the detector must be optimized for **R**, it is natural to measure **I** also through optical (UV) signals. **R** is prompt may be considered primary scintillation, but not in the usual sense, and is detected within about 50 ns of the recoil event. The **I** signal is sensed after drift, with delays ranges from about one μs to hundreds.
R/I may display a soft dependence on total energy, as well as (by design) a strong dependence on CR. The total energy **E** of the event is a weighted sum: **E = aR + bI**. The coefficient "a" will depend on the spatial variation of detection efficiency of light within the active volume. This spatial variation may be modest, since optical detection is foreseen to closely approach 4π, as elaborated below. Both coefficients will differ substantially depending on whether the event is an electron or nuclear recoil.

3. The constraint on density: the columnarity of events

Density enters prominently in the determination of optimum operating conditions. Taking r_O again as the Onsager radius and defining d as measure of nuclear recoil track range, the aspect ratio $A = d/r_O$ provides a useful geometrical measure of "electrostatic columnarity". If the average distance between single ionization events were equal to the Onsager radius, the atomic density could be described as "matched" to the average ionization density. In an average sense, "Onsager spheres" then connect the track segments continuously, *i.e.*, as in a string of beads. In this idealized case of averaged continuous connectivity of the Onsager spheres, the recoil track attains a maximum length.
The optimum for sensing directionality of recoil, however, may not be this idealized case. The atomic density for which directional sensitivity is optimized may correspond to an A value smaller than the simplest case above. A substantial level of Onsager sphere overlap may be necessary to obtain retention and focusing of the electron cloud by the collective positive charge of the ion column. Presently, this is one of the uncertainties needing exploration.
For example, a 30 keV nuclear recoil in xenon gas for $\rho = 0.05$ g/cm^3 (about ten bars) has a range of about 2100 nm (see below). The aspect ratio $A = d/r_O$ is approximately 30 in this case. With $r_O = 70$ nm in xenon gas, the minimum dE/dx for Onsager spheres to "touch" is ~14 electron/ion pairs per μm. As discussed in the next section, the maximum ionization density of 30 keV nuclear recoils in xenon gas at 0.05 g/cm^3 (about ten bars) falls in the range of 100 electron-ion pairs/μm, about seven times the geometric *minimum* dE/dx. In this case of substantial average overlap, an "Onsager tube" of larger radius is the more appropriate image. Ambipolar diffusion is present. The aspect ratio A is more difficult to estimate, but is still above 5. Clearly the optimum density is unlikely to lie much above $\rho =$ 0.05 g/cm^3, about 1 ½ % of LXe density.

For comparison, the range in LXe for a 30 keV nuclear recoil track is about 35 nm. $A = d/r_O < 1$, and overlap of Onsager spheres is complete. No electrostatic sense of directionality remains in LXe for nuclear recoils of interest, and directional sensing is beyond hope.

4. What are the conditions for a useful sensitivity?

The necessary conditions appear to be five:

1. WIMPs have masses roughly similar to that of xenon; low mass WIMPs transfer too little energy, and very high mass WIMPs become excessively rare.
2. Nuclear recoils retain, in a statistical sense, significant directionality relative to the incoming WIMP direction.
3. The spatial structure of the primary ions and electrons created after the nuclear recoil event collectively retain a sufficiently linear character;
4. The population of atoms/molecules that experience columnar recombination in a typical event is large enough that useful statistical precision exists;
5. The population of atoms/molecules that experience columnar recombination can be measured with adequate precision and accuracy.

The first condition remains unknown, but is not disfavored theoretically. The code Stopping and Range in Matter (SRIM) provides an approximate answer for condition 2. The author of SRIM does not guarantee accuracy for results obtained for gas densities. Without recourse to a more accurate method, I show results from SRIM in figure 2 for 30 keV xenon atoms, impinging from the left. The recoiling nucleus moves sufficiently slowly, $\sim\frac{2}{3} \times 10^{-3}$ c, that the atom itself is the recoiling object; the atom becomes ionized as it collides with other xenon atoms. Scattering is substantial, but the ions largely retain their original directions during their initial trajectories – where energy loss is greatest. The average energy lost to ionization is 6.1 keV, reflecting the predominant loss of energy to heat, and a corresponding quenching factor of about 5.

Another shortcoming of SRIM for present purposes is that no information about excitations is available. No experimental data for energy transfer by nuclear recoils to ionization and excitation are available for the gaseous phase of xenon. Liquid and gas phases may display quite different behaviors due to the presence of a conduction band in LXe, absent in the gas phase at $\rho = 0.05$ g/cm^3. Nevertheless, it is reasonable to assume that the slowly moving atomic recoils produce many more excitations than ionizations. Primary excitations, however, add nothing to directional sensitivity, and represent an important source of recoverable energy that should not be ignored. The stratagem for conversion to additional ionization is elaborated in the next section.

For condition 3, the high density of newly created ions retains the cloud of electrons for a time sufficient for thermalization and subsequent recombination of electrons, under the assumption that the imposed electric field is not too high. For condition four, three variables exist for the optimization of CR: density of xenon, electric field, and fractional addition of a molecular component – easy to accomplish in gas phase xenon. For condition five, the use of a (possibly unique) gaseous molecular additive –trimethylamine – appears to offer the possibility to maximize optical detection efficiency.

5. Trimethylamine – a Penning molecule with ideal fluorescence

Xenon with trimethylamine (TMA) has been shown to serve as a strong Penning mixture, although the evidence is indirect [3]. A xenon-TMA Penning mixture will convert the energy harbored in the primary excitations to ionization. The newly created ionization can contribute to a CR signal. The presence of TMA also hastens electron thermalization, which in pure xenon is very slow.

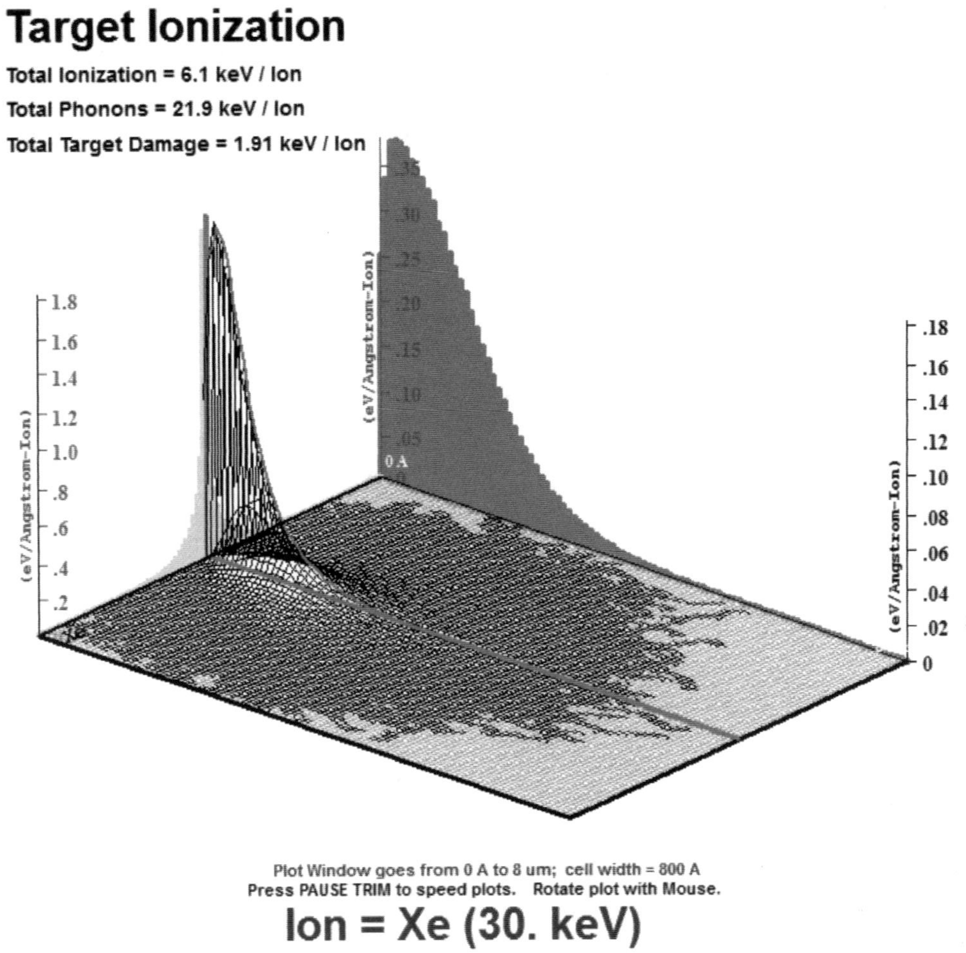

Figure 2. This SRIM plot shows how energy is distributed in an 8 x 8 μm^2 plane for 100,000 xenon ions of 30 keV entering from the left along the X-axis, the red line. The red histogram displays the deposited ionization along the original ion direction, X, integrated over Y and Z. The green histogram shows the transverse distribution of ionization – note the sharp peak at Y = 0. The grey histogram indicates ionization density in the plane. The xenon gas density is ρ = 0.05 g/cm^3, about ten bars. An energy weighted range is about ~2 μm for these conditions.

TMA is also one of the three tertiary amines studied for fluorescence [4]. When excited by photons of ~240 nm, TMA returns fluorescence with unitary efficiency in a band centered at 300 nm. Providentially, this band matches the peak of the excitation spectrum for common commercial wavelength shifting plastic (WLS). Thus the addition of TMA to the xenon gas permits the TPC interior to be fully covered (4π) with WLS. Earlier work with triethylamine and argon mixtures showed that electroluminescence is possible without charge gain. This is expected to be the case with TMA + xenon, since the ionization potentials of TMA and TEA are nearly identical, and electrons only reach energies corresponding to excitation levels of TEA/TMA.

This design concept is illustrated in figure 3, showing that only a few dozen PMTs are needed to detect the light captured within the WLS bars. The 4π coverage also softens spatial dependence of the optical detection efficiency. An overall optical detection efficiency of 10% may be achievable, improving the statistical precision of the **R** and **I** signals. The untrapped fraction of WLS light emerging from the end-cap WLS plates provides sufficient intensity for an array of SiPMs to provide the tracking function.

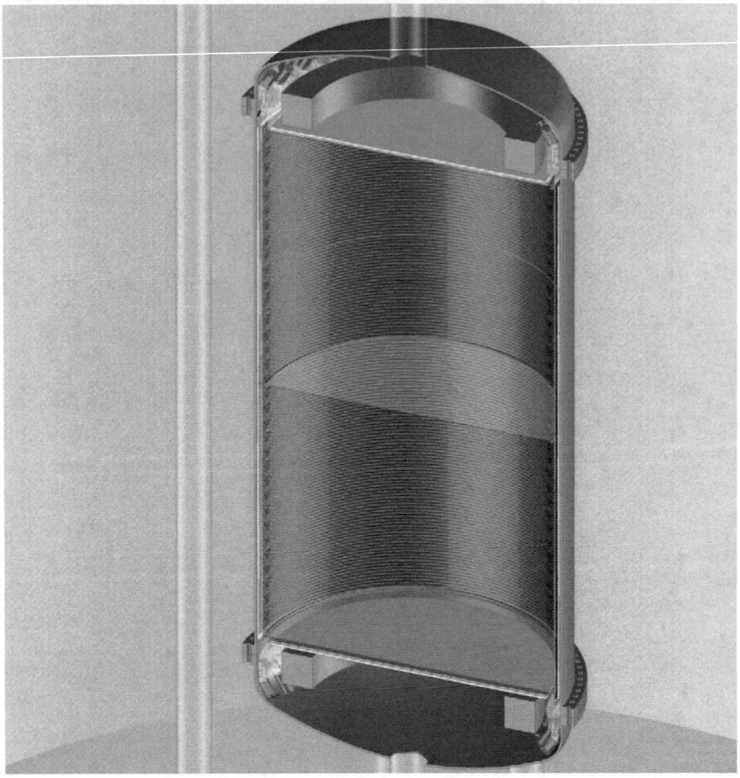

Figure 3. This figure shows a symmetric TPC design for ton-scale active mass at pressure of 10 bars. The width of the TPC is 2.4 m, and each drift distance is 2.5 m. WLS plastic covers the end (anode) regions as well as the walls of the cylindrical field cage to achieve 4π coverage. Tracking with small SiPMs is still achievable, using the untrapped light emerging from the WLS end-plates. The PMTs are hidden behind a copper shield, minimizing their contribution to radioactivity budget. Although unknown at present, a head-tail effect may appear as a difference in **R/I** between the upper and lower halves of the TPC.

6. Discussion

The detector could work anywhere, from polar to equatorial regions, by orienting to compensate for the angle of the earth's rotation axis. It is necessary to orient the detector such that the drift field is aligned with the expected WIMP flow once per sidereal day. Ideally, the detector should also be anti-aligned once per sidereal day to minimize systematic effects, unfortunately, not possible with a single, fixed, single-ended TPC detector at any latitude. It is possible, however, to construct a symmetric TPC, for which ½ has the field aligned parallel once a day, and, simultaneously the the other ½ is anti-parallel. Then the modulation difference between the two halves would automatically display a head-tail effect statistically.

Although α particles differ from nuclear recoils in important ways, such as their ability to produce numerous δ-rays up to about 1 keV – which low-energy nuclear recoils cannot produce, and a range about 100x greater, their use could help to understand the optimum conditions to achieve an optimal directionality sensitivity exploiting CR.

At present, this concept includes several uncertain features, most of which can be addressed experimentally and by simulations. The experimental uncertainties include:

1. Do the nuclear recoils retain a sufficient level of their initial direction, while dissipating most of their kinetic energy?
2. Does TMA deliver a useful level of free ionization by Penning transfer, and what is the optimum fraction that maximizes the Penning transfer to free ionization?
3. What fraction of TMA provides optimum thermalization of the electrons?
4. What is the dependence of optimum electric field as a function of xenon density?

Journal of Physics: Conference Series **460** (2013) 012006 doi:10.1088/1742-6596/460/1/012006

5. What is the dependence of optimum TMA fraction with xenon density?
6. Does the charge exchange process of xenon with TMA occur rapidly enough to convert the ion image to TMA ions?

Despite the number of uncertain features in this concept, it seems prudent and valuable to prepare for the eventuality that direct detection experiments provide evidence supporting the WIMP nature of dark matter. A robust signature of directional anisotropy in WIMP-nuclear recoils seems the essential element for a claim of discovery. I gratefully acknowledge many useful discussions and contributions to this concept by my collaborators in NEXT-100.

References

[1] arXiv:0911.0323v1 [astro-ph.CO] 1 Nov 2009
[2] G C Jaffe, "Zur Theorie der Ionisation in Kolonnen", Annalen der Physik, Vol. 42, p. 303, 1913.
[3] B D Ramsey, P C Agrawal, "Xenon-based Penning Mixtures for Proportional Counters", Nuc. Inst. & Meth. **A278** 1989 p576-582
[4] C G Cureton *et al,* "The Photophysics of Tertiary Aliphatic Amines", Chem. Phys. **63** (1981) p 31-49

Journal of Physics: Conference Series **460** (2013) 012007 doi:10.1088/1742-6596/460/1/012007

MIMAC: A micro-tpc matrix for dark matter directional detection

D. Santos, J. Billard[1], G. Bosson, J.L. Bouly, O. Bourrion, C. Fourel, O. Guillaudin, J. Lamblin, J.F. Muraz, F. Mayet, J.P. Richer, Q. Riffard

LPSC, Universite Joseph Fourier Grenoble 1, CNRS/IN2P3, Institut Polytechnique de Grenoble

[1] Present address: Department of Physics and Kavli Institute for Astrophysics and Space Research, Massachusetts Institute of Technology, Cambridge, MA 02139, USA,

E. Ferrer, I. Giomataris, F.J. Iguaz[2]

IRFU,CEA Saclay, 91191 Gif-sur-Yvette cedex

[2] Present address: Laboratorio de Física Nuclear y Astropartículas, Universidad de Zaragoza, Spain

L. Lebreton, D. Maire

LMDN, IRSN Cadarache, 13115 Saint-Paul-Lez-Durance

E-mail: `Daniel.Santos@lpsc.in2p3.fr`

Abstract.

The dark matter directional detection opens a new field in cosmology bringing the possibility to build a map of nuclear recoils that would be able to explore the galactic dark matter halo giving access to a particle characterization of such matter and the shape of the halo. The MIMAC (MIcro-tpc MAtrix of Chambers) collaboration has developed in the last years an original prototype detector based on the direct coupling of large pixelized micromegas with a devoted fast self-triggered electronics showing the feasibility of a new generation of directional detectors. The discovery potential of this search strategy is discussed and illustrated. In June 2012, the first bi-chamber prototype has been installed at Modane Underground Laboratory (LSM) and the first underground background events, the gain stability and calibration are shown.

1. Introduction

Cosmological and astrophysical observations converge to a standard cosmological model requiring a new kind of particle. The most favourable candidate is a stable weakly interacting massive particle (WIMP). In the context of direct detection of these new particles an alternative, complementary and in any case needed strategy is the development of detectors providing an unambiguous positive WIMP signal. Indeed, directional detection gives a new degree of freedom to discriminate statistically the WIMP interactions from neutrons, the particles that produce the same expected signal, giving in addition a clear signature for events produced by particles in our galactic halo. This can be achieved by searching for a correlation of the WIMP signal with the solar motion around the galactic centre, observed as a direction dependence of the WIMP stream [1], coming from (l = 90°, b = 0°) in galactic coordinates, which happens to be roughly in the direction of the constellation Cygnus. The background events, coming from gamma rays and neutrons produced in the atmosphere or in the rock should follow the rotation of the Earth, isotropic in galactic coordinates and easy to discriminate from those coming from the Cygnus constellation direction. A dedicated statistical study with simulated data analysis has shown that even a low-exposure, directional detector could allow a high significance discovery of galactic Dark Matter even with a background contamination or to a robust and competitive exclusion curve [5], depending on the value of the unknown WIMP-nucleon cross section. In [7], a study has been performed on the capability of directional detectors to probe neutralino dark matter in the Minimal Supersymmetric Standard Model and the Next-to-Minimal Supersymmetric Standard Model with parameters defined at the weak scale. It shows that directional detectors such as the future MIMAC detector (50 m^3) will probe spin dependent dark matter scattering on nucleons that are beyond the reach of current spin independent detectors since the scalar and axial cross section are not correlated [7].

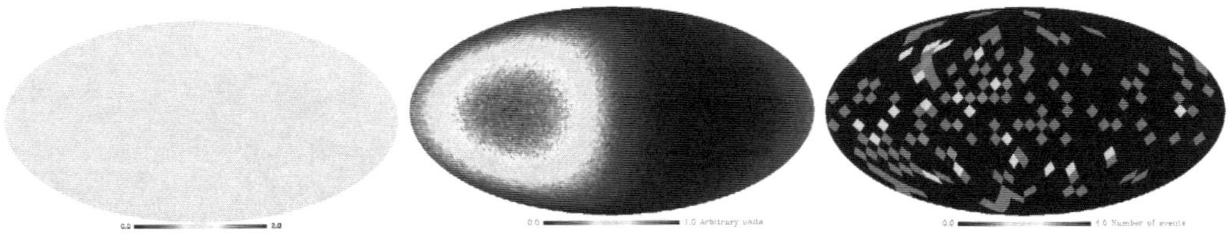

Figure 1. From left to right: isotropic background distribution, WIMP-induced recoil distribution in the case of an isothermal spherical halo and a typical simulated measurement : 100 WIMP-induced recoils and 100 background events with a low angular resolution. Recoils maps are produced for a ^{19}F target, a 100 GeV.c^{-2} WIMP and considering recoil energies in the range 5 keV $\leq E_R \leq$ 50 keV. Figures from [4].

The right panel of figure 1 presents a typical recoil distribution observed by a directional detector : 100 WIMP-induced events and 100 background events generated isotropically. For an elastic axial cross-section on nucleon $\sigma_{\rm n} = 1.5 \times 10^{-3}$ pb and a 100 GeV.c^{-2} WIMP mass, this corresponds to an exposure of $\sim 1.6 \times 10^3$ kg.day in CF_4, on their equivalent energy ranges as discussed in ref. [4]. Low resolution maps are used in this case ($N_{\rm pixels} = 768$) which is sufficient for the low angular resolution, $\sim 15°$ (FWHM), expected for this type of detector. In this case, 3D readout and sense recognition are considered, while background rejection is based on electron/recoil discrimination by track length and energy selection [12]. In order to conclude from the recoil map of figure 1 (right) that it does contain a fraction of WIMP events pointing towards the direction of the solar motion a likelihood analysis has been developed. The likelihood value is estimated using a binned map of the overall sky with Poisson statistics, as shown in [4]. This is a four parameter likelihood analysis with m_χ, $\lambda = S/(B+S)$ the WIMP fraction (B is the background spatial distribution taken as isotropic and S is the WIMP-induced recoil distribution) and the coordinates (ℓ, b) referring to the maximum of the WIMP event angular distribution. The result of this map-based likelihood method is that the main recoil direction is recovered and is pointing towards ($\ell = 95° \pm 10°, b = -6° \pm 10°$) at 68 % CL, corresponding to a non-ambiguous detection of particles from the galactic halo. This is indeed the discovery proof of this detection strategy [4]. As emphasized in ref. [2], a directional detector could allow for

Journal of Physics: Conference Series **460** (2013) 012007 doi:10.1088/1742-6596/460/1/012007

a high significance discovery of galactic Dark Matter even with a small amount of background contamination. This holds true even when astrophysical and experimental uncertainties are taking into account. For very low exposures, competitive exclusion limits may also be imposed [5].

There are many projects around the world [18], [19], [20], trying to show the ability to get the directionality at low nuclear recoil energies summarized in [21]. The MIMAC (MIcro-tpc MAtrix of Chambers) detector project [17] tries to get these elusive events by a double detection: recoil ionization and track, at low gas pressure with low mass target nuclei (H, ^{19}F or ^{3}He). In order to have a significant cross section we explore the axial, spin dependent, interaction on odd nuclei. The very weak correlation between the neutralino-nucleon scalar cross section and the axial one, as it was shown in [6], [7] makes this research, at the same time, complementary to the massive target experiments.

2. The MIMAC bi-chamber prototype

The MIMAC bi-chamber prototype consists of two chambers of (10 cm x 10 cm x 25 cm) with a common cathode, which is an elementary module of the future matrix. The purpose of this prototype is to show the ionization and track measurement performances needed to achieve the directional detection strategy. The primary electron-ion pairs produced by a nuclear recoil in one chamber of the matrix are detected by driving the electrons to the grid of a bulk micromegas [8] and producing the avalanche in a very thin gap (256μm).

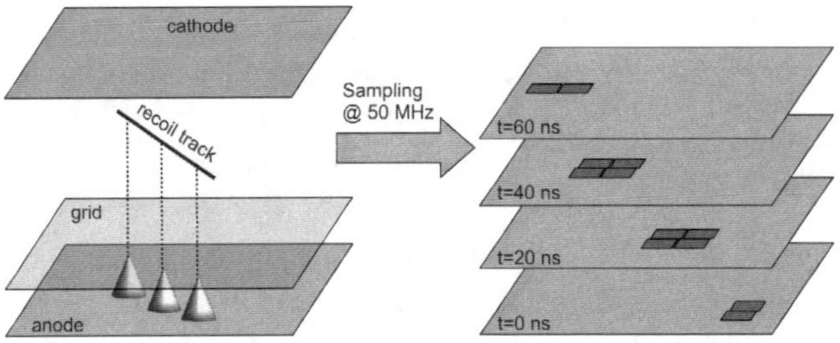

Figure 2. The anode is read every 20 ns and knowing the drift velocity of primary electrons, the 3D track can be reconstructed from the consecutive number of images defining the event.

As pictured on figure 2, the electrons are collected towards the grid in the drift space and are multiplied by avalanche to the pixellized anode thus allowing to get information on X and Y coordinates. To have access to the X and Y coordinates a bulk micromegas with a 10 by 10 cm active area, segmented in pixels with a pitch of 424 μm was used as 2D readout [9]. In order to reconstruct the third coordinate Z of the points of the recoil track, the LPSC developed a self-triggered electronics able to perform the anode sampling at a frequency of 50 MHz. This includes a dedicated 64 channels ASIC [15] associated to a DAQ [16].

In order to get the total recoil energy we need to know the ionization quenching factor (IQF) of the nuclear recoil in the gas used. We have developed at the LPSC a dedicated experimental facility to measure such IQF. A precise assessment of the available ionization energy has been performed in ^{4}He + 5%C_4H_{10} mixture within the dark matter energy range (between 1 and 50 keV) by a measurement of the IQF [10]. For a given energy, an electron track in a low pressure micro-TPC is an order of magnitude longer and showing more straggling than a recoil one. It opens the possibility to discriminate electrons from nuclei recoils by using both energy and track information, as it was shown in [11] and [12].

3. Reconstruction of 3D particle tracks

The first result concerning the ability to detect tracks with the prototype was performed with a ^{55}Fe X-ray source in order to reconstruct the 5.9 keV electron tracks produced by photoelectric effect in the active volume. The 3D tracks are obtained from consecutive read-outs of the anode, every 20 ns, defining the event. To get the length and the orientation of the track, a statistical treatment of maximization of the likelihood with the MIMAC observables and an independent measurement of the drift velocity is needed [3]. Measurements of the drift velocity have been performed [14], and they fit well with Magboltz simulations [13].

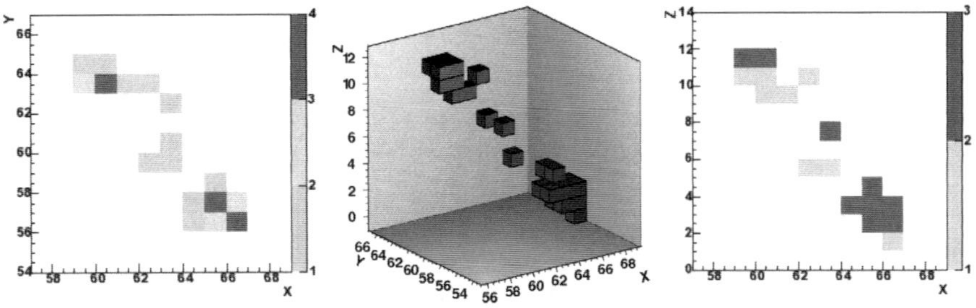

Figure 3. A 5.9 keV electron track in 350 mbar 95% $^4He + C_4H_{10}$. The left panel represents the 2D projection of the recoil seen by the anode, the center panel represents a 3D view of the track after using the reconstruction algorithm and the right panel represents a projection of the 3D track on the XZ plane

Figure 3 presents a typical electron track seen by the anode (X,Y) (left panel), its projection on the XZ plane (right panel) and reconstructed in 3D (center panel). This result shows the MIMAC capability to reconstruct the track of low energy electrons at high gain which are the typical background in dark matter experiments.

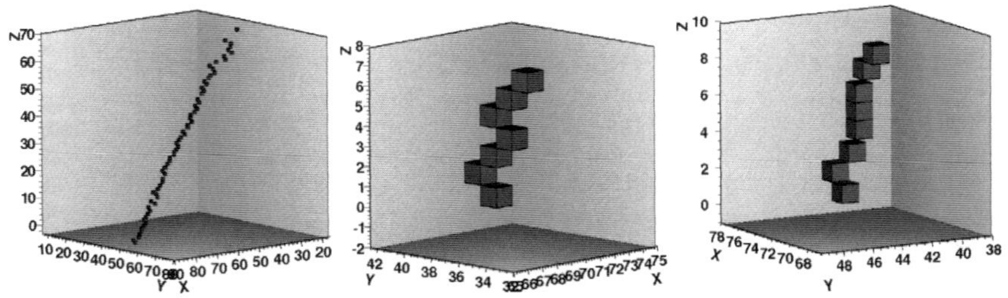

Figure 4. From left to right: 3D track reconstructions, from the barycenters of the anode strips of pixels read every 20 ns for : a 5.5 MeV alpha particle in 350 mbar of $^4He + 5\%C_4H_{10}$, a 8 keV hydrogen nucleus in 350 mbar of $^4He + 5\%C_4H_{10}$ and a fluorine nucleus leaving 50 keV in ionization in 55 mbar 70% CF_4 + 30% CHF_3

On fig 4 (left panel), a 3D track reconstruction is achieved for high energy (5.5 MeV) alpha particles issued from the natural radioactivity (^{222}Rn) present in the chamber, from the barycenters of the anode strips of pixels read every 20 ns. However, the final validation concerning the possibility for MIMAC to get directional detection had to be done with neutrons giving nuclear recoils in the range of a few keV. In order to have mono-energetic neutron fields, in the range of a few tens of keV, we have performed an experiment at the AMANDE facility

(IRSN- Cadarache) allowing to select the energy of the neutrons by the angle with respect to a proton beam producing a neutron resonance on a LiF target.

On fig 4 (center and right panel), 3D tracks of nuclear recoils following elastic scattering of mono-energetic neutrons are represented. On the center panel, a 8 keV proton recoil leaving a track of 2.4 mm long in 350 mbar of $^4He + 5\% C_4H_{10}$ is represented. The right panel presents a 50 keV (in ionization) fluorine recoil of 3 mm long obtained in a 55 mbar mixture of 70% CF_4 + 30% CHF_3.

4. MIMAC at the Underground Laboratory of Modane (LSM)

In June 2012, we have installed the bi-chamber module, at the Underground Laboratory of Modane (LSM), see fig.5 .

Figure 5. The bi-chamber module installed at Modane in June 2012.

The set-up includes a close loop gas system with in-line filtering and is able to reach a vacuum of 10^{-6} mbar. The pressure was regulated at 50 mbar. The gas mixture used was 70% CF_4 + 28% CHF_3 + 2% C_4H_{10} We have calibrated twice a week by means of fluorescence produced by X generator on thin foils of Cd (3.2 keV), Fe (6.4 keV), Cu (8.1 keV) and Pb (10.5 and 12 keV). In the fig. 6, we show the low energy calibration obtained, its linearity and the stability of the calibration given by the bin position of the different peaks as a function of time. The first data acquisition started on June 22nd and it has been continuously run and remotely monitored up to October 12th .

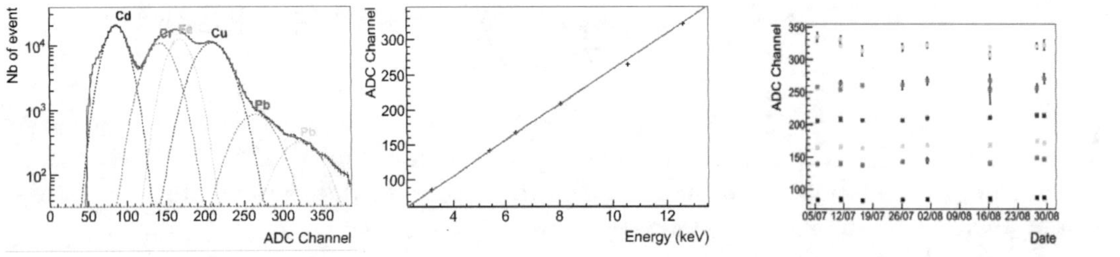

Figure 6. Left: The fit of the five X-ray peaks produced by fluorescence. Centre: The linear calibration obtained from the fit. Right: the gain stability from July 5th to August 28th.

In order to characterize the total background of our detector at Modane, we worked without any shielding. Besides the very good stability of the calibration validating the gas circulation, one of the most interesting facts observed during this first run was the determination of the main sources of ^{222}Rn inside the gas circuit.

The 10 cm by 10 cm micromegas coupled to the MIMAC electronics running the 512 channels per chamber and working at high gain and without any problem during a long period of time (4 months in our first run at Modane), is in fact the validation of the feasibility of a large TPC for directional detection. As an illustration of the quality of data obtained at Modane, we show on fig.7 a very interesting recoil event of 34 keV in ionization. The analysis of this first run at Modane is in progress.

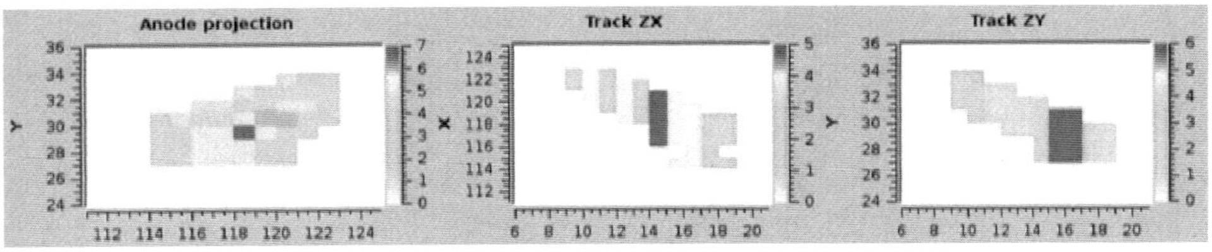

Figure 7. On the left, the X-Y plane of the anode showing the intersections of the strips fired. On the centre, the X projection as a function of time, every 20 ns. On the right, the same but for the Y projection

5. Conclusions

Directional detection is a promising search strategy to discover galactic dark matter. The MIMAC detector provides the energy of a recoiling nucleus and the reconstruction of its 3D track. The first 3D tracks observed with the MIMAC prototype were shown: 5.9 keV electrons and low energy proton and fluorine recoils.The installation of the first bi-chamber prototype at Modane was succesfully done in June 2012. The stability of the gain was characterized and the main source of background events was determined. The next step will be to build a demonstrator of 1 m^3 to show that a large micro-tpc matrix for directional detection of dark matter search is possible.

6. References

[1] D. N. Spergel, Phys. Rev. D **37** (1988) 1353.
[2] J. Billard, F. Mayet and D. Santos, Phys. Rev. D **85** (2012) 035006
[3] J. Billard *et al.*, JCAP **04** (2012) 006.
[4] J. Billard *et al.*, Phys. Lett. B **691** (2010) 156-162.
[5] J. Billard *et al.*, Phys. Rev. D **82** (2010) 055011.
[6] E. Moulin, F. Mayet and D. Santos. Phys. Lett. B **614** (2005) 143-154.
[7] D. Albornoz Vazquez *et al.* Phys. Rev. D **85** (2012) 055023.
[8] I. Giomataris *et al.*, Nucl. Instrum. Methods A **560** (2006) 405.
[9] F. Iguaz *et al.*, JINST 6 (2011) P07002
[10] D. Santos *et al.*, arXiv:0810.1137.
[11] D. Santos *et al.* , CYGNUS 2011, EAS Publications Series 53 (2012) 25-31.
[12] J. Billard *et al.*, JCAP 07 (2012) 020.
[13] S.F. Biagi, Nucl. Instrum. and Meth. A**421** (1999) 234-240.
[14] J. Billard *et al.*, in preparation.
[15] J.P. Richer *et al.*, Nucl. Instrum. Methods A **620** (2010) 470.
[16] O. Bourrion *et al.*,Nucl. Instrum. and Meth. A**662**(2010) 207.
[17] D. Santos *et al.*, J. Phys. Conf. Ser. **65** (2007) 012012.
[18] G. J. Alner *et al.*, Nucl. Instr. Meth. A **555** (2005) 173.
[19] S. Ahlen *et al.*, arXiv:1006.2928.
[20] K. Miuchi *et al.*, Phys. Lett. B **654** (2007) 58.
[21] S. Ahlen *et al.*, Int. J. Mod. Phys. A **25** (2010) 1.

Journal of Physics: Conference Series **460** (2013) 012008 doi:10.1088/1742-6596/460/1/012008

Status report on the GBAR CERN experiment

Pascal Debu

Irfu, CEA Saclay, 91191 Gif-sur-Yvette CEDEX, France

E-mail: pascal.debu@cea.fr

Abstract. GBAR is a recently approved experiment at CERN/AD. Its aim is to perform the first test of the Weak Equivalence Principle with Antimatter. The objective is to measure the gravitational acceleration of antihydrogen atoms on Earth with 1 % precision in a first phase, and better than a per mil in a second phase. The method is to detect the free fall of ultra-cold atoms. By sympathetic cooling of antihydrogen positive ions $\overline{\mathrm{H}}^+$ with beryllium ions, and after photodetachment of the excess positron, antiatoms are formed at m/s velocities and their free fall can be directly measured. Antihydrogen ions are produced by the interaction of keV antiprotons with a dense positronium cloud, the latter being formed by dumping positrons onto a porous material. We describe here briefly the experimental techniques and report on the most recent results. In particular, a flux of around $4 \times 10^6\,\mathrm{e^+s^{-1}}$ slow positrons is now produced by the source installed at Saclay, cross sections for the production of the $\overline{\mathrm{H}}^+$ ions have been computed, and a dedicated beam line for the study of positronium formation and for applications in Materials Science has been realized. The tentative schedule of the GBAR developments is given.

1. Introduction

The GBAR experiment has been approved by the CERN Research Board in May 2012 [1]. Its aim is to perform the first test of the Weak Equivalence Principle (WEP) with antimatter. The Einstein Equivalence Principle is at the heart of general relativity, it has been tested with a very high precision with matter, but there is no conclusive direct measurement with antimatter. This is a basic scientific question, the interest of which is enhanced by the unknown origin of the acceleration of the expansion of the universe and by the hypothetical presence of dominant quantities of dark matter: these observations suggest that our understanding of gravitation may be incomplete.

Indirect tests of the Equivalence Principle for antimatter can be obtained by comparing the properties of particle-antiparticle systems. Two particle-antiparticle systems have provided such limits: the $K^0 - \bar{K}^0$ system (comparison of decay parameters [2]), and the $p - \bar{p}$ system (simultaneous measurement of both cyclotron frequencies [3]). One can also use the WEP tests with matter: because of the matter antimatter virtual pair content of the nuclei, if antimatter is coupled differently to gravitation from matter, this induces differences between various nuclei. However, as shown in [4], all these tests rely upon debatable theoretical hypotheses.

A limit on the mass difference between neutrinos and antineutrinos has been derived by the comparison of the time arrival of events from the 1987A supernova explosion [5], but this limit relies on the non proven hypothesis that at least one out of the 19 observed events is due to an electron neutrino, the other ones being antineutrino interactions.

Previous attempts to test the WEP with positrons [6] and antiprotons [7] have been launched, but neither has been completed. It seems also out of present reach to perform gravity experiments with antineutrons [8] or positronium [9]. The antihydrogen atom is consequently the best candidate system.

2. Principle and goal of the experiment

GBAR is described in detail in [10]. The key innovative feature is to produce and cool to around 10 µK $\overline{\mathrm{H}}^+$ ions to provide ultra-cold neutral $\overline{\mathrm{H}}$ atoms by photodetachment of the extra positron. The duration time of their subsequent free fall down a few tens of cm provides a direct measurement of their gravity acceleration ($\bar{g}$). This method has been proposed by Walz and Hnsch in [11], but the production of enough $\overline{\mathrm{H}}^+$ ions to perform the experiment was not addressed. The main source of uncertainty comes from the initial speed of the anti-atom, and the precision on $\bar{g}$ is mainly statistical. About 1500 ions are needed to reach a 1 % relative precision. This is the goal of the first phase of the experiment.
$\overline{\mathrm{H}}^+$ ions can be produced by two charge-exchange processes: $\bar{\mathrm{p}} + \mathrm{Ps} \rightarrow \overline{\mathrm{H}} + \mathrm{e}^-$ followed by $\overline{\mathrm{H}} + \mathrm{Ps} \rightarrow \overline{\mathrm{H}}^+ + \mathrm{e}^-$. Because the cross sections of these reactions are very low, the experiment relies on the production of large quantities of low energy (in the keV range) antiprotons, and on a very high flux of positrons, well above the capacity of ^{22}Na sources, in order to produce enough Ps. The overall scheme of the measurement is as follows:

- Electrons of 10-20 MeV from a small linear accelerator are dumped onto a tungsten target and produce fast (MeV) positrons.
- The emitted positrons are moderated down to 3 eV and accumulated in a Penning-Malmberg trap.
- Once a few 10^{10} positrons are stored, they are ejected onto a positron-positronium converter.
- a laser pulse allows the excitation of a fraction of the Ps to enhance the $\overline{\mathrm{H}}^+$ production yield ;
- In coincidence, 100 keV antiprotons are extracted from the antiproton decelerator AD followed by the ELENA ring presently in construction at CERN ; they are decelerated further down to a few keV and directed towards the newly formed dense positronium cloud.
- The $\overline{\mathrm{H}}^+$ ions produced are selected, decelerated and trapped in a segmented Paul trap, where they are sympathetically cooled with beryllium ions.
- The excess positron of $\overline{\mathrm{H}}^+$ is photodetached at threshold and the free fall duration time is measured by detecting the antihydrogen atom annihilation on a plate below the trap with time projection chambers and scintillating counters. Because of the unknown initial velocity, $\bar{\mathrm{g}}$ has to be statistically extracted.

3. Slow positron production

A prototype slow positron source is installed at IRFU at Saclay. It comprises a 4.3 MeV electron linear accelerator producing bursts of 2.5 µs effective length at 200 Hz with 140 µA mean current, a 1 mm tungsten target, and a moderator made of a stack of tungsten meshes. Figure 1 shows the e^+ yield for various moderator setups. The optimization is still under way. The present achieved positron rate is in the range 3-4$\times 10^6\,\mathrm{s}^{-1}$. It is comparable to rates obtained with the most intense ^{22}Na radioactive sources. The target rate for GBAR is $3 \times 10^8\,\mathrm{s}^{-1}$. It will be reached by increasing the electron accelerator energy and repetition rate, and by improving the moderation technique.

4. Positron accumulation

Slow positrons are guided through a vacuum pipe by a 8 mT magnetic field to a Multi Ring Trap provided by the RIKEN atomic laboratory, and described in [12]. With a 3 T magnetic field in

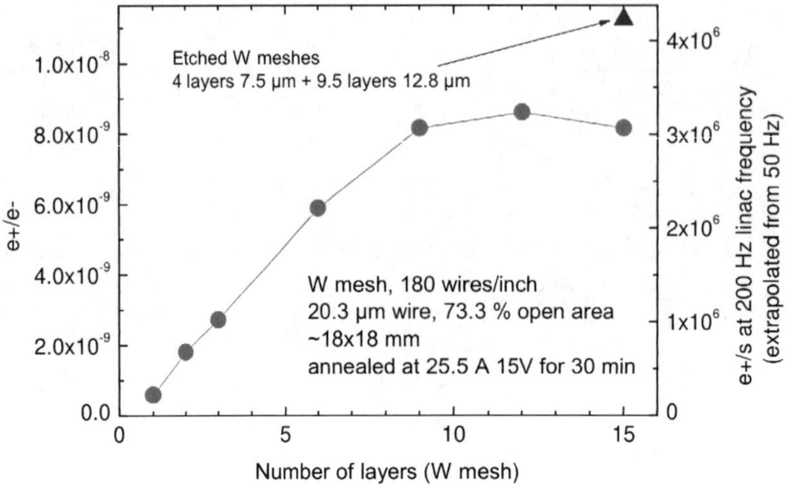

Figure 1. Positron yield as a function of the number of W meshes layers. Annealing of W meshes at high temperature is necessary to obtain a good moderation efficiency. Chemical etching can also improve the yield as shown by the blue triangle. The diameter of the mesh wires is another relevant parameter.

the trap, the transport efficiency into the trap has been measured to be close to 100 % when an acceleration voltage of 1.6 kV is applied at the entrance of the trap to pass the magnetic mirror. The trapping scheme will take advantage of the pulsed slow positron beam: positrons are injected in the trap by lowering the voltage of the entrance electrode ; the potential is restored before positrons make a round trip. They are slowed down by Coulomb interactions with a preloaded electron plasma, and are eventually trapped in a potential well in less than 3 ms, before the next burst arrives and the entrance electrode is lowered again. Because it takes 80 ns for positrons to make a round trip in the trap, it is necessary to bunch the positron beam. The present method is to apply an increasing voltage during the burst so that late positrons are faster than early ones. Because of the energy and angular dispersion of the positrons emitted by the moderator, the bunching efficiency is only around 30 %. A dedicated buncher will be necessary to improve this efficiency. The setup is now ready to perform systematic studies of the trapping mechanism.

5. Production of a dense positronium target.

Orthopositronium (oPs), which has a lifetime of 142 ns in its ground state, is produced by dumping the positrons accumulated in the trap in less than 100 ns on a converter material, the one presently used is the nanoporous SiO_2. The production efficiency has been measured to be as high as 30 % [13, 14, 15]. To further study and optimize the positronium production, a secondary beamline has been assembled end of 2012 (Figure 2). This line will also be used for applications in Materials Science.

Figure 2. Photograph of the experimental setup at Saclay. One sees on the left the main positron beam line with the Riken trap, and on the right the secondary beam line.

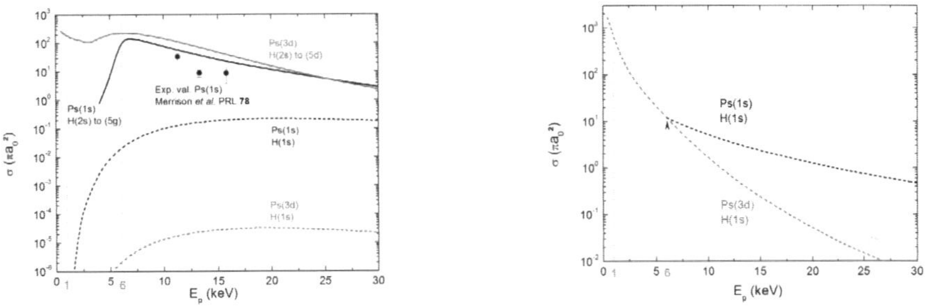

Figure 3. Cross-section in atomic units for the reactions $\overline{\mathrm{p}} + \mathrm{Ps}^{(*)} \rightarrow \overline{\mathrm{H}}^{(*)} + \mathrm{e}^-$ (left) and $\overline{\mathrm{H}} + \mathrm{Ps}^{(*)} \rightarrow \overline{\mathrm{H}}^+ + \mathrm{e}^-$ (right) as a function of the antiproton kinetic energy

6. Production of antihydrogen positive ions

In coincidence with the formation of the positronium cloud, a bunch of 5×10^7 antiprotons is directed to cross that cloud. $\overline{\mathrm{H}}^+$ ions are produced via the two-step process: $\overline{\mathrm{p}} + \mathrm{Ps} \rightarrow \overline{\mathrm{H}} + \mathrm{e}^-$ followed by $\overline{\mathrm{H}} + \mathrm{Ps} \rightarrow \overline{\mathrm{H}}^+ + \mathrm{e}^-$. The cross section of the matter counterpart of the first reaction has been measured above 10 keV and estimated over a wider energy range [16]. It is of the order of 10^{-15} cm^2, with a strong variation with energy. The second reaction has not been experimentally verified. In [17], the estimate for its matter counterpart is about 10^{-16} cm^2. However, because the binding energy of the n = 3 positronium state is 0.75 eV and is very close to that of $\overline{\mathrm{H}}^+$, the cross section is expected to be strongly enhanced with excited positronium. The cross section of these processes have been recently evaluated for different excited states of the antihydrogen atoms and of the positronium [18]. The cross section increase is strongly dependent on the incident energy (Figure 3). The optimization of the whole formation process is under way, it involves the optimization of the fraction of Ps to be excited, the choice of the antiproton kinetic energy, the duration of the laser pulse, and the length of the reaction chamber.

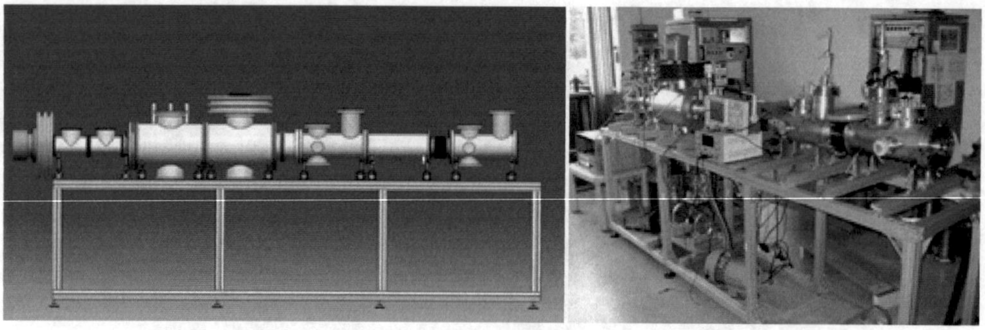

Figure 4. Schematic (left) and photograph (right) of the decelerator test bench in Orsay.

First estimates show that it is reasonable to expect one $\overline{\mathrm{H}}^+$ per delivered antiproton bunch. The Ps production and excitation will be tested at CEA Saclay in 2014.
To maximize the $\overline{\mathrm{H}}^+$ yield, antiproton kinetic energy must be in the range 1 to 6 keV. This is much lower than the expected 100 keV energy which will be provided by the ELENA ring. It will thus be necessary to further decelerate the antiprotons. This will be done with an electrostatic decelerator tube. This method has been experienced at ISOLDE on 60 keV ions. Very good efficiencies can be obtained (see for example [19]). With the updated parameters of ELENA, a preliminary estimate of the efficiency to transport the decelerated antiproton beam to the reaction chamber is around 40 %. The decelerator is being realized at Orsay CSNSM for tests with protons (Figure 4).

7. Cooling of the antihydrogen ions

The $\overline{\mathrm{H}}^+$ ions have almost the same energy as the incident antiprotons. They have to be selected and slowed down to enter a segmented Paul trap where they will undergo sympathetic cooling with Be^+ ions. The method is described in detail in the proposal. It has been shown that Be^+ ions can be cooled to temperatures below 10 µK [21]. However, the sympathetic cooling of $\overline{\mathrm{H}}^+$ ions to less than 10 µK is a double challenge : sympathetic cooling followed by side-band cooling in a Paul trap with such a large charge/mass difference between the ion species has not been demonstrated, and it will the first time that one attempts this technique with antimatter, for which losses of antiatoms must be avoided. Detailed simulations are under way, and the technique will be developed by LKB and Mainz University with hydrogen ions before installation at CERN.

8. Free fall detection

To minimize the size of the ultra-vacuum chamber, the free fall of the $\overline{\mathrm{H}}$ atom will be performed on a few tens of cm height. It will last a few tenths of a second (assuming $\bar{g}$ = g !). The photodetachment of the bound positron from the ion has to be done very close to threshold to avoid a too large recoil which would prevent making the measurement with a detector of reasonable size. The first recoil is due to the absorption of the photon, it is harmless since the recoil velocity is 0.2 ms^{-1}, and the direction can be chosen to be horizontal. The second recoil is due to the emitted positron. If the photon is tuned to below 10 µeV above threshold, the recoil velocity remains below 1 ms^{-1}. The cross section of the photodetachment will be very low, but the laser power can be focused on a very small volume, because the ions can be localized to around 10 µm. From measurements [22, 23], one gets $\sigma \approx 4 \times 10^{-16}(\Delta\mathrm{E[eV]})^{3/2}\mathrm{cm}^2$, leading to a cross section of order 10^{-23} cm^2 at 10 µeV above threshold. With a 1 W laser beam focused

on a 100 μm^2 area, the photodetachment rate is around 100 s^{-1} per ion, more than enough to perform all measurements during the delay between two antiproton ejections from the AD (110 s). The laser shot time will give the start time of the free fall.
The antihydrogen annihilation on a detector plane at a distance h from the Paul trap center will be detected and the duration Δt of the free fall measured, leading to an estimate of the acceleration of antihydrogen: $\bar{g} = 2h/\Delta t^2$, assuming a null mean initial velocity of the antiatoms. The detector will consist in a set of Time Projection Chambers covering as much as possible of the vacuum chamber to reconstruct the vertex position of the antihydrogen annihilation and reject the cosmic background, and scintillator counters surrounding the whole setup to measure the annihilation time.
By putting all parameters together, one meets the requirements of the method proposed in [11] and a 1 % precision on $\bar{g}$ is reachable in a few weeks of running time, enough to accumulate a few thousands detected $\overline{H}$ ions.

9. Perspectives

It will be possible to reach a much higher precision by using the Casimir effect in the same way as was demonstrated with ultra-cold neutrons [24] : if the detector plane is a few micron below the Paul trap centre from where the ultra-cold $\overline{H}$ atoms are released, those are reflected with very high probability by the detector plane. They are trapped by the gravitational field. Their height is quantized and the energy levels are proportional to the gravitational mass, allowing a measurement of $\bar{g}$ to 10^{-3} relative precision with a few hundreds events.

10. Collaboration and tentative schedule

The GBAR collaboration gathers 48 physicists from 14 institutes covering the many fields needed to perform the experiment : accelerator, particle and atomic physics, plasma trapping, positronium and materials science, antihydrogen formation, laser cooling and atom manipulation, gravitational spectrometry : CSNSM Orsay, ETH Zurich, ILL Grenoble, IPCMS Strasbourg, IRFU Saclay, Lebedev Moscow, LKB Paris, JGU Mainz, NCBJ Otwock Swierk, RIKEN, Swansea University, University of Tokyo Komaba, Tokyo University of Science, Uppsala University.
GBAR was approved by CERN Research Board in May 2012. Tests of trapping positrons have started at Irfu Saclay. The deceleration technique will be tested in 2013 at CSNSM Orsay. The positronium formation and excitation will be measured in 2014. The free fall detector is expected to be ready for tests by 2015. The sympathetic cooling must be first demonstrated with ordinary matter ions by LKB and Mainz University before the installation at CERN. A new electron accelerator will be built at NCBJ in Poland by 2015. Commissionning of the experiment should start in 2016. The physics run is scheduled for 2017.

Acknowledgements

In addition to the institutes and laboratories involved in GBAR, part of this work has been funded by the "Conseil Général de l'Essonne" and by the "Agence Nationale de la Recherche" in France.

References

[1] CERN Research Board 2012 Minutes of the 200th Meeting of The Research Board of 30 May 2012 *Preprint* CERN/DG/Research Board 2012-429
[2] CP LEAR collaboration 1999 Tests of the Equivalence Principle with neutral kaons *Phys. Lett.* B **452** 425
[3] Gabrielse G *et al* 1999 Precision Mass Spectroscopy of the Antiproton and Proton Using Simultaneously Trapped Particles *Phys. Rev. Lett.* **82** 3198
[4] Nieto M M and Goldman T 1991 The Arguments Against Antigravity and the Gravitational Acceleration of Antimatter *Phys. Rep.* **205** 221

[5] Pakvasa S, W Simmons W and Weiler T 1989 Test of equivalence principle for neutrinos and antineutrinos *Phys. Rev.* D **39** 1761
[6] Witteborn F and Fairbank W 1967 Experimental Comparison of the Gravitational Force on Freely Falling Electrons and Metallic Electrons *Phys. Rev. Lett.* **19** 1049
[7] Beverini N *et al* 1986 PS200 proposal *Preprint* Los Alamos National Laboratory report LAUR-86-260
[8] Brando T 1981 Observations of low-energy antineutrons in a time-separated neutral beam, *Nucl. Instr. Methods* **180** 461
[9] Mills A P 2002 Positronium molecule formation, Bose - Einstein condensation and stimulated annihilation *Nucl. Instr. Methods in Physics Research* B **192** 107
[10] Chardin G *et al* 2011 Proposal to measure the Gravitational Behaviour of Antihydrogen at Rest GBAR *Preprint* CERN-SPSC-P-342
[11] Walz J and Hnsch T 2004 A Proposal to Measure Antimatter Gravity Using Ultracold Antihydrogen Atoms *General Relativity and Gravitation* **36** 561
[12] Oshima N *et al* 2004 New Scheme for Positron Accumulation in Ultrahigh Vacuum *Phys. Rev. Lett.* **93** 195001
[13] Liszkay L *et al* 2008 Positronium reemission yield from mesostructured silica films *Appl. Phys. Lett.* **92** 063114
[14] Crivelli P *et al* Measurement of the ortho-positronium confinement energy in mesoporous thin films *Phys. Rev.* A **81** 052703
[15] Cassidy D B *et al* 2010 Positronium cooling in porous silica measured via Doppler spectroscopy *Phys. Rev.* A **81** 012715
[16] Merrison J P *et al* 1997 Hydrogen Formation by Proton Impact on Positronium, *Phys. Rev. Lett.* **78** 2728
[17] Walters H R J and Starrett C 2007 Positron and positronium scattering *Phys. Stat. Sol.* C 3429
[18] Comini P and Hervieux P A 2012 *Journal of Physics: Conference Series* 16th International Conference on Positron Annihilation to be published
[19] Lunney D *et al* 2009 COLETTE a linear Paul-trap beam cooler for the on-line mass spectrometer MISTRAL *Nucl. Instr. Meth. in Physics Research* A **598** 379
[20] Eriksson T *et al* 2010 ELENA - An Updated Cost and Feasibility Study *Preprint* CERN-BE-2010-29 OP
[21] Barrett M D *et al* 2003 Sympathetic cooling of $^9Be^+$ and $^24Mg^+$ for quantum logic *Phys. Rev.* A **68** 042302
[22] Smith S J and Burch D S 1959 Photodetachment cross section of the negative hydrogen ion *Phys. Rev. Lett.* **2** 165
[23] Lykke K R , Murray K K and W C Lineberg W C 1991 Threshold photodetachment of H^- *Phys. Rev.* A **43** 6104
[24] Nesvizhevsky V V 2002 *et al* Quantum states of neutrons in the Earth's gravitational field *Nature* **415** 297

Journal of Physics: Conference Series **460** (2013) 012009 doi:10.1088/1742-6596/460/1/012009

Search for neutrino-less double beta decay with EXO-200 and prospects for Ba ion tagging

R Gornea[1]
Physics Professor, *Laboratory for High Energy Physics*,
Albert Einstein Center for Fundamental Physics, Sidlerstrasse 5, Bern, CH

E-mail: gornea@zoho.com

Abstract. The Enriched Xenon Observatory collaboration searches for neutrino-less double beta decay in ^{136}Xe. The first phase of the experiment, EXO-200 has started taking data in 2011 and has already produced important physics results like the first observation of the allowed two-neutrino double beta decay in ^{136}Xe. A massive effort for material screening has accompanied the design of EXO-200 and has produced a detector with very low background due to the residual radioactivity. The most recent data analysis campaign has provided a strong limit for the neutrino-less double beta decay in ^{136}Xe. For the next phase ton-scale detector, the collaboration is investigating the options for employing a Ba ion tagging technique to provide complete suppression of the radioactive background.

References

The material screening efforts in preparation of EXO-200 are described in [1]. Various critical components of EXO-200 are described in [2,3,4,5]. A complete review of EXO-200 design and construction is given in [6]. The most recent analysis of EXO-200 data is presented extensively in [8].

[1] Leonard D S *et al.* 2008 *Systematic study of trace radioactive impurities in candidate construction materials for EXO-200* Nucl. Instr. Meth. **A** 591 p 490
[2] Neilson R *et al.* 2009 *Characterization of large area APDs for the EXO-200 detector* Nucl. Instr. Meth. **A** 608 p 68
[3] Dobi A *et al.* 2011 *A xenon gas purity monitor for EXO* Nucl. Instr. Meth. **A** 659 p 215
[4] LePort F *et al.* 2011 *A magnetically-driven piston pump for ultra-clean applications* Rev. Sci. Instr. 82 p 105114
[5] Dobi A *et al.* 2012 *Xenon purity analysis for EXO-200 via mass spectrometry* Nucl. Instr. Meth. **A** 675 p 40
[6] Auger M *et al.* 2012 *The EXO-200 detector, part I: detector design and construction* J. Instr. 7 p 05010
[7] Ackerman N *et al.* 2011 *Observation of two-neutrino double-beta decay in ^{136}Xe with the EXO-200 detector* Phys. Rev. Lett. 107 p 212501
[8] Auger M *et al.* 2012 *Search for Neutrinoless Double-Beta Decay in ^{136}Xe with EXO-200* Phys. Rev. Lett. 109 p 032505

[1] To whom any correspondence should be addressed.

Journal of Physics: Conference Series **460** (2013) 012010 doi:10.1088/1742-6596/460/1/012010

Status and physics potential of NEXT-100

J. Martín-Albo and J.J. Gómez-Cadenas,
on behalf of the NEXT Collaboration

Instituto de Física Corpuscular (IFIC), CSIC & Universidad de Valencia
Calle Catedrático José Beltrán, 2, 46980 Paterna, Valencia, Spain

E-mail: justo.martin-albo@ific.uv.es, gomez@mail.cern.ch

Abstract. The NEXT-100 time projection chamber, currently under construction, will search for neutrinoless double beta decay ($\beta\beta0\nu$) using 100–150 kg of high-pressure xenon gas enriched in the ^{136}Xe isotope to $\sim 91\%$. The detector possesses two important features for $\beta\beta0\nu$ searches: very good energy resolution (better than 1% FWHM at the Q value of ^{136}Xe) and event topological information for the distinction between signal and background. Furthermore, the technique can be extrapolated to the ton-scale, thus allowing the full exploration of the inverted hierarchy of neutrino masses.

1. Introduction

Neutrinoless double beta decay ($\beta\beta0\nu$) is a postulated very slow nuclear transition in which two neutrons undergo β decay simultaneously and without the emission of neutrinos [1, 2]. The importance of this process can hardly be overstated. An unequivocal observation would establish that massive neutrinos are Majorana particles [3, 4] — that is, identical to their antiparticles —, implying that a new physics scale beyond the Standard Model must exist [5, 6]. Furthermore, it would prove that total lepton number is not conserved, suggesting that this new physics could also be the reason for the observed asymmetry between matter and antimatter in the universe [7, 8].

Many mediating mechanisms — involving, in general, physics beyond the Standard Model — have been proposed for $\beta\beta0\nu$, the simplest one being the virtual exchange of light Majorana neutrinos [1]. Assuming this to be the dominant mechanism at low energies, the decay rate of $\beta\beta0\nu$ can be written as [2]

$$(T^{0\nu}_{1/2})^{-1} = \frac{F_N}{m_e}\, m^2_{\beta\beta}\,. \tag{1}$$

In this equation, F_N is a nuclear structure factor that depends on the particular isotope that is disintegrating, m_e is the mass of the electron, and $m_{\beta\beta}$ is the *effective Majorana mass* of the electron neutrino:

$$m_{\beta\beta} = \left|\sum_i U^2_{ei}\, m_i\right|, \tag{2}$$

where m_i are the neutrino mass eigenstates and U_{ei} are elements of the neutrino mixing matrix. In consequence, a measurement of the $\beta\beta0\nu$ decay rate would provide direct information on neutrino masses [1].

2. Experimental searches for neutrinoless double beta decay

The detectors used to search for $\beta\beta0\nu$ are designed, in general, to measure the energy of the radiation emitted by a $\beta\beta0\nu$ source. In a neutrinoless double beta decay, the sum of the kinetic energies of the two released electrons is always the same, and is equal to the mass difference between the parent and the daughter nuclei: $Q_{\beta\beta} \equiv M(Z, A) - M(Z+2, A)$. However, due to the finite energy resolution of any detector, $\beta\beta0\nu$ events would be reconstructed within a given energy range centered around $Q_{\beta\beta}$ and typically following a gaussian distribution. Other processes occurring in the detector can fall in that region of energies, thus becoming a background and compromising drastically the sensitivity of the experiment [9].

All double beta decay experiments have to deal with an intrinsic background, the standard two-neutrino double beta decay ($\beta\beta2\nu$), that can only be suppressed by means of good energy resolution. Backgrounds of cosmogenic origin force the underground operation of the detectors. Natural radioactivity emanating from the detector materials and surroundings can easily overwhelm the signal peak, and hence careful selection of radiopure materials is essential. Additional experimental signatures that allow the distinction of signal and background are a bonus to provide a robust result.

Besides energy resolution and control of backgrounds, several other factors such as detection efficiency and scalability to large masses must be taken into consideration in the design of a double beta decay experiment. The simultaneous optimization of all these parameters is most of the time conflicting, if not impossible, and consequently many different experimental techniques have been proposed. In order to compare them, a figure of merit, the experimental sensitivity to $m_{\beta\beta}$, is normally used [9]:

$$m_{\beta\beta} \propto \sqrt{1/\varepsilon}\left(\frac{b\,\delta E}{M\,t}\right)^{1/4}, \tag{3}$$

where ε is the signal detection efficiency, M is the $\beta\beta$ isotope mass used in the experiment, t is the data-taking time, δE is the energy resolution and b is the specific background rate in the region of interest around $Q_{\beta\beta}$ (expressed in counts per kg of $\beta\beta$ isotope, year and keV).

Until very recently, the experimental searches for $\beta\beta0\nu$ had been dominated by germanium calorimeters, mainly due to their excellent energy resolution. In particular, for about a decade the best limit to the half-life of $\beta\beta0\nu$ was the one set by the Heidelberg-Moscow (HM) experiment: $T^{0\nu}_{1/2}(^{76}\mathrm{Ge}) \geq 1.9 \times 10^{25}$ years at 90% CL [10]. A subgroup of this experiment interpreted the same data as evidence of a positive signal, with a best value for the half-life of 1.5×10^{25} years, or 0.39 eV in terms of $m_{\beta\beta}$ [11]. This claim has been practically ruled out by two new experiments searching for $\beta\beta0\nu$ in ^{136}Xe, EXO-200 [12, 13] and KamLAND-Zen [14, 15], that published their first results during 2012. Their basic experimental parameters, as defined in equation (3), are collected in Table 1. The combination of the limits reported by the two experiments gives $T^{0\nu}_{1/2}(^{136}\mathrm{Xe}) > 3.4 \times 10^{25}$ years (90% CL), refuting the HM claim at > 97.5% CL [15].

Xenon is indeed an interesting species for double beta decay searches. Two of its natural isotopes, ^{134}Xe and ^{136}Xe, are $\beta\beta0\nu$-decaying candidates. The latter, having a higher $Q_{\beta\beta}$ value (2458 keV [16]), is preferred for $\beta\beta0\nu$ searches because the decay rate is proportional to $Q^5_{\beta\beta}$ and the radioactive backgrounds are less abundant at higher energies. The $\beta\beta2\nu$ mode of ^{136}Xe ($\sim 2.3 \times 10^{21}$ years) is slow compared to that of other sources, and therefore the experimental requirement for energy resolution is less stringent. Moreover, the process of isotopic enrichment is relatively simple and less costly than for the other $\beta\beta0\nu$ isotopes, and consequently ^{136}Xe-based experiments are the most obvious candidates for a future multi-ton experiment.

The KamLAND-Zen experiment is a modification of the well-known KamLAND neutrino detector [17]. A transparent balloon, ~ 3 m diameter, containing 13 tons of liquid scintillator loaded with 330 kg of enriched xenon (91% of ^{136}Xe) is suspended at the center of KamLAND. The scintillation light generated by events occurring in the detector is recorded by an array of

Table 1. Basic experimental parameters of the three ^{136}Xe-based double beta decay experiments: mass of ^{136}Xe, M; signal detection efficiency, ε; energy resolution at the Q value of ^{136}Xe, δE; and background rate, b, in the region of interest around $Q_{\beta\beta}$ expressed in counts/(keV kg year) (shortened as ckky). The parameters of EXO-200 and KamLAND-Zen are those published by the collaborations after the initial operation of the detectors [13, 15], while those of NEXT come from results obtained with the prototypes and predictions based on simulations.

Experiment	M (kg)	ε (%)	δE (% FWHM)	b (10^{-3} ckky)
EXO-200	90	62	3.9	1.5
KamLAND-Zen	300	42	9.9	1.0
NEXT-100	90	25	0.7	0.5

photomultipliers surrounding it. From the detected light pattern, the position of the event vertex is reconstructed with a spatial resolution of about 15 cm/$\sqrt{E(\text{MeV})}$. The energy resolution is $(6.6 \pm 0.3)\%/\sqrt{E(\text{MeV})}$, that is, 9.9% FWHM at the Q value of ^{136}Xe. The signal detection efficiency is 42% due to the tight fiducial cut introduced to reject backgrounds originating in the balloon. The achieved background rate in the energy window between 2.2 MeV and 3.0 MeV is 0.001 counts/(keV kg year). The dominant source of background is ^{100m}Ag, a β^- emitter. Its presence in the detector may be due to contamination of the balloon by Fukushima fallout during fabrication or cosmogenic production by Xe spallation.

EXO-200, unlike KamLAND-Zen, uses the enriched xenon as both source and detection medium. The detector is a cylindrical time projection chamber (TPC) filled with 110 kg of liquid xenon (LXe) enriched to 81% in ^{136}Xe. The interaction of charged particles with the LXe produces both a scintillation and an ionization signal. The TPC is divided into two symmetric halves separated by a cathode grid, and each end is instrumented with a pair of crossed wire planes that detect the ionization, and behind those, an array of 250 APDs is used to record the scintillation light. The energy resolution obtained combining the two signals is 3.9% FWHM at $Q_{\beta\beta}$. The topological reconstruction of the events is used for discriminating signal and background. The background rate in the region of interest around $Q_{\beta\beta}$ is 1.5×10^{-3} counts/(keV kg year). The dominant sources of backgrounds are the external high-energy gammas and the radioactive contaminants in the detector vessel.

3. NEXT-100, high-pressure xenon gas for $\beta\beta0\nu$ searches

The NEXT-100[1] detector [18] will search for the neutrinoless double beta decay of ^{136}Xe using a time projection chamber filled with 100–150 kg of enriched xenon gas at 10–15 bar pressure. Such a detector offers both very good energy resolution (better than 1% FWHM at 2.5 MeV) and event topological information to discriminate between signal and background. As we will see, this combination results in excellent sensitivity to $\beta\beta0\nu$. In addition, the technology can be extrapolated to a ton-scale experiment, thus allowing the full exploration of the inverted hierarchy of neutrino masses [1]. The detector is currently under construction, and its installation and commissioning at the *Laboratorio Subterráneo de Canfranc* (LSC), in Spain, are planned for the first semester of 2014.

During the last three years, the NEXT Collaboration has developed an R&D program with the specific goal of proving the performance of the technology. This program has resulted in the construction and operation of the NEXT-DEMO [19] and NEXT-DBDM [20] prototypes. The first results of these detectors have been presented at this conference [21].

[1] NEXT stands for *Neutrino Experiment with a Xenon gas Time projection chamber.*

Gaseous xenon, as a detection medium, provides, like LXe, *scintillation* and *ionization* as primary signals. The former is used in NEXT to establish the start-of-event time (t_0), while the latter is used for calorimetry and tracking. In its gaseous phase, xenon can provide high energy resolution, in principle as good as 0.3% FHWM at the Q value of ^{136}Xe [22]. In order to achieve optimal energy resolution, the ionization signal is amplified in NEXT using the electroluminescence (EL) of xenon: the electrons liberated by ionizing particles passing through the gas are first drifted towards the TPC anode by a weak electric field ($\sim$ 300 V/cm), entering then into another region where they are accelerated by a high electric field ($\sim$ 25 kV/cm at 10 bar), intense enough so that the electrons can excite the xenon atoms but not enough to ionize them. This excitation energy is ultimately released in the form of proportional (with sub-poissonian fluctuations) secondary scintillation light.

NEXT-100 will have different readout systems for calorimetry and tracking. An array of 60 photomultiplier tubes (Hamamatsu R11410-10 PMTs), the so-called *energy plane*, located behind the TPC cathode detects a fraction of the secondary scintillation light to provide a precise measurement of the total energy deposited in the gas. These PMTs detect as well the primary scintillation, used to signal the start of the event. The forward-going secondary scintillation is detected by a dense array of 1-mm^2 silicon photomultipliers (SiPMs), known as the *tracking plane*, located behind the anode, very close to the EL region, and is used for event topological reconstruction.

The R11410-10 PMT was specially developed for radiopure, xenon-based detectors. However, pressure-resistance tests run by the manufacturer showed that this PMT cannot withstand pressures above 6 atmospheres. Therefore in NEXT-100 they will be sealed into individual pressure resistant, vacuum tight copper enclosures coupled to sapphire windows (see figure 1, left). The chosen SiPM for NEXT-100 is the S10362-11-050P model by Hamamatsu. This device has an active area of 1 mm^2, 400 sensitive cells (50 μm size) and high photon detection efficiency in the blue region (about 50% at 440 nm). The dark count rate is 0.4 MHz, that is, less than 1 event per microsecond (which is the considered sampling time). This random noise events have amplitudes of up to 8 photoelectrons, and thus a digital threshold at those levels should lead to an insignificant noise rate in NEXT-100 without affecting the tracking performance. The SiPMs will be mounted in cuflon or cirlex boards (depending on the radiopurity of these materials) spaced 1 cm (see figure 1, right). The photon detection efficiency (PDE) of the chosen SiPMs peaks in the blue region of the spectrum, and they are not sensitive below 200 nm, where the emission spectrum of xenon lies. Consequently, the boards with the SiPMs will be coated with tetraphenyl butadiene (TPB), a wavelength shifter [23].

The active volume of NEXT-100 — a cylinder 130 cm long and 105 cm diameter — can hold 100 kg of xenon gas at 15 bar, and it is surrounded by a copper shell 15 cm thick that shields it against external backgrounds. The TPC and the detector planes are housed in a stainless-steel (316Ti alloy) pressure vessel. A lead castle made of 20 cm thick bricks attenuates the external flux of high-energy gammas emanating from the laboratory walls. A drawing of the detector and shielding is shown in figure 2.

4. Projected sensitivity of the NEXT-100 experiment

NEXT has two powerful handles to distinguish signal from background:

- *Energy resolution*: Signal events have all the same energy. Selecting only the events in the energy region around $Q_{\beta\beta}$ defined by the resolution eliminates most of the spurious activity in the detector.
- *Event topology*: Electron tracks in high-pressure xenon (HPXe) are tortuous due to multiple scattering, and have a distinctive energy deposition pattern with a roughly-constant $\mathrm{d}E/\mathrm{d}x$ except at the end, where it increases by a factor of 4 to 5, leaving a large deposition that

Journal of Physics: Conference Series **460** (2013) 012010 doi:10.1088/1742-6596/460/1/012010

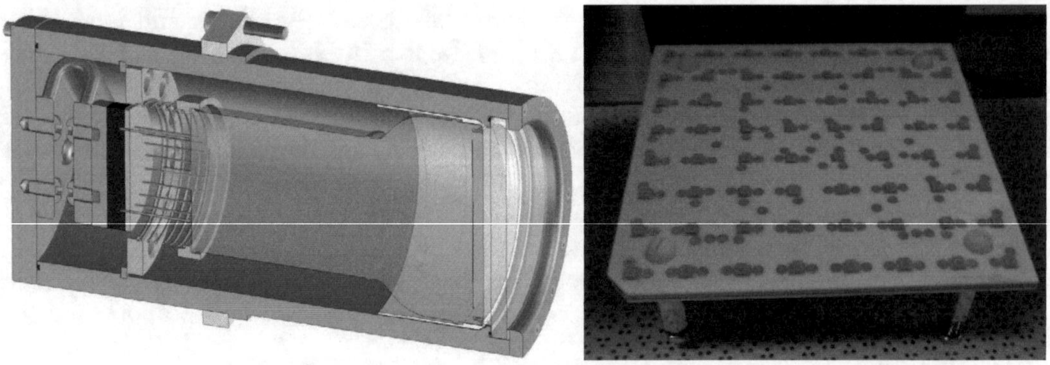

Figure 1. Left: Drawing of a Hamamatsu R11410-10 PMT inside its pressure-resistant copper enclosure. Right: A *dice board* containing 64 (8 × 8) SiPMs illuminated with a UV lamp after being coated with TPB.

Figure 2. Cross-section drawing of the NEXT-100 TPC inside its lead shielding.

we call a *blob*. This topological signature can be used in the search for $\beta\beta0\nu$ to distinguish between signal (two electron tracks with a common vertex) and background (mostly single electrons originated in the interaction of high-energy gammas), as demonstrated by the Gotthard experiment [24].

The relevance of a background source depends therefore on its probability of generating a signal-like track in the active volume with energy around $Q_{\beta\beta}$. The $\beta\beta0\nu$ peak of ^{136}Xe is located between the photo-electric peaks of the high-energy, de-excitation gammas emitted following the β-decays of ^{214}Bi, from the uranium series, and ^{208}Tl, from the thorium series. Most materials contain impurities of those isotopes in a given amount, and thus a thorough radiopurity control

Table 2. Estimated contribution of several subsystems to the total background rate of the NEXT-100 detector, assuming for the PMTs the conservative level of background (5 mBq/PMT).

Subsystem	^{208}Tl (10^{-4} ckky)	^{214}Bi (10^{-4} ckky)
Pressure vessel	0.1	0.1
Field cage	0.2	0.2
Energy plane	2.2	0.9
Tracking plane	1.0	0.2
Total	3.5	1.4

is being performed for the construction of the detector. Measurements of activity levels in the most relevant materials have been carried out based on GDMS and on ultra-low background germanium gamma-ray spectrometry at the LSC Radiopurity Service [25].

Of special importance are the photomultipliers, which may be the dominant source of background in NEXT-100. Currently, there are only upper limits to their background level. The most sensitive measurement, performed by the LUX collaboration, sets a limit in the background of each PMT of less than 700 μBq [26], while the XENON collaboration quotes a less sensitive limit of about 5 mBq/PMT [27]. The NEXT Collaboration plans to screen the 60 PMTs to be installed in the detector, but for the moment we consider the quoted measurements as the optimistic and conservative limits, respectively, of the contribution of the PMTs to the total background rate.

Table 2 summarizes the contributions of the main subsystems of NEXT-100 to the total background rate, computed using a detailed simulation of the detector and the measurements from the radiopurity campaign. A conservative background rate of 5×10^{-4} counts/(keV kg year) is obtained when the XENON limit is used for the background level of the PMTs, whereas a rate of about 10^{-4} counts/(keV kg year) is obtained for the LUX limit. Taking into account the conservative background estimate, and the computed signal efficiency (see table 1), we have evaluated the sensitivity of the NEXT-100 experiment and compare it to that of KamLAND-Zen and EXO-200, assuming that the experimental parameters quoted in table 1 do not change significantly in the following years. The result is shown in figure 3. The curve corresponding to the NEXT-100 experiment drops faster than those of the other two experiments, due to better energy resolution and background suppression. This compensates its late start. By 2018 the three experiments will reach a similar sensitivity of about 130 meV. By 2020, NEXT-100 will reach, after a 5-years run, a sensitivity of about 100 meV.

5. Conclusions

Neutrinoless double beta decay experiments are the only practical way to establish whether neutrinos are Majorana particles. The current generation of experiments, with a sensitivity to $m_{\beta\beta}$ of about 100 meV, is already operating, with initial results from EXO-200 and KamLAND-Zen, both using ^{136}Xe. The NEXT-100 detector could contribute in a decisive way to this exploration, in spite of a relatively late start.

At the same time, exploring the inverse hierarchy will require large masses (circa 1 ton) and a background level in the regime of very few counts per ton-year. Many of the current technologies may be eliminated either by lack of resolution or by the difficulties to extrapolate to large masses. In contrast, the NEXT concept which combines very good resolution, a topological signature and a clean detector, could lead the way in the near future.

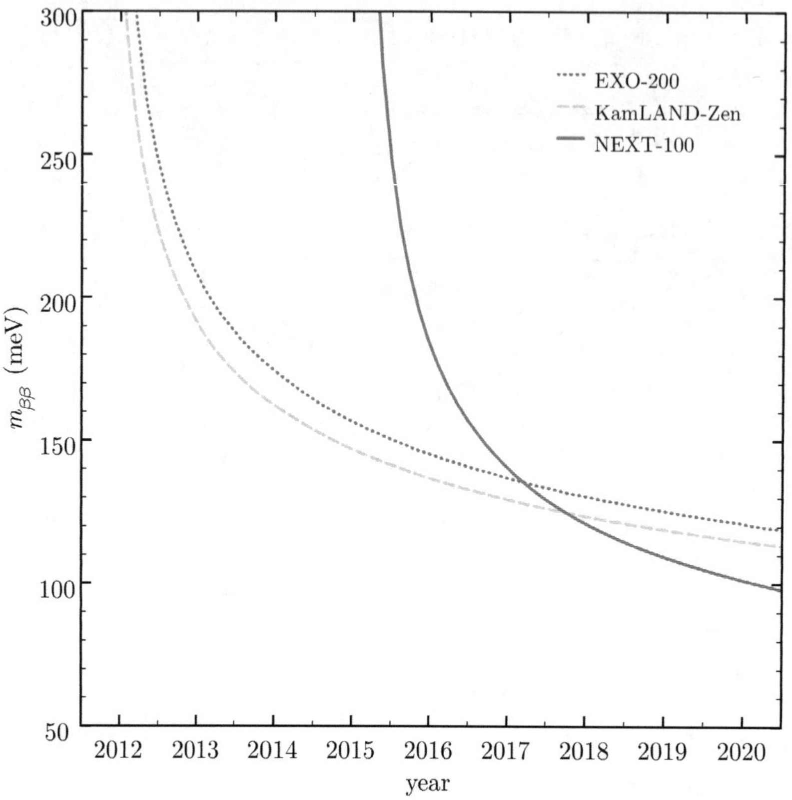

Figure 3. Projected sensitivity of NEXT-100 (continuous, blue line) in a five-years run (2015–2020) compared with the sensitivities of the other two ^{136}Xe-based experiments, KamLAND-Zen (dashed, green line) and EXO-200 (dotted, magenta line). The sensitivities have been calculated assuming the parameters in Table 1 and a 10% deadtime for the three experiments.

Acknowledgments
This work was supported by the *Ministerio de Economía y Competitividad* of Spain under grants CONSOLIDER-Ingenio 2010 CSD2008-0037 (CUP) and FPA2009-13697-C04-04.

References

[1] Gómez-Cadenas J J, Martín-Albo J, Mezzetto M, Monrabal F and Sorel M 2012 *Riv. Nuovo Cim.* **35** 29–98 (*Preprint* 1109.5515)
[2] Avignone III F T, Elliott S R and Engel J 2008 *Rev. Mod. Phys.* **80** 481–516 (*Preprint* 0708.1033)
[3] Furry W H 1939 *Phys. Rev.* **56** 1184–1193
[4] Schechter J and Valle J W F 1980 *Phys. Rev. D* **22** 2227
[5] Beringer J *et al.* (Particle Data Group) 2012 *Phys. Rev. D* **86** 010001
[6] González-García M C and Maltoni M 2008 *Phys. Rept.* **460** 1–129 (*Preprint* 0704.1800)
[7] Fukugita M and Yanagida T 1986 *Phys. Lett. B* **174** 45
[8] Davidson S, Nardi E and Nir Y 2008 *Phys.Rept.* **466** 105–177 (*Preprint* 0802.2962)
[9] Gómez-Cadenas J J, Martín-Albo J, Sorel M, Ferrario P, Monrabal F *et al.* 2011 *JCAP* **1106** 007 (*Preprint* 1010.5112)
[10] Klapdor-Kleingrothaus H V *et al.* 2001 *Eur. Phys. J. A* **12** 147–154 (*Preprint* hep-ph/0103062)
[11] Klapdor-Kleingrothaus H V, Dietz A, Harney H L and Krivosheina I V 2001 *Mod. Phys. Lett. A* **16** 2409–2420 (*Preprint* hep-ph/0201231)

[12] Auger M *et al.* (EXO Collaboration) 2012 *JINST* **7** P05010 (*Preprint* 1202.2192)
[13] Auger M *et al.* (EXO Collaboration) 2012 *Phys. Rev. Lett.* **109** 032505 (*Preprint* 1205.5608)
[14] Gando A *et al.* (KamLAND-Zen Collaboration) 2012 *Phys. Rev. C* **85** 045504 (*Preprint* 1201.4664)
[15] Gando A *et al.* (KamLAND-Zen Collaboration) 2012 (*Preprint* 1211.3863)
[16] McCowan P and Barber R 2010 *Phys. Rev. C* **82** 024603
[17] Abe S *et al.* (KamLAND Collaboration) 2010 *Phys. Rev. C* **81** 025807 (*Preprint* 0907.0066)
[18] Álvarez V *et al.* (NEXT Collaboration) 2012 *JINST* **7** T06001 (*Preprint* 1202.0721)
[19] Álvarez V *et al.* (NEXT Collaboration) 2012 (*Preprint* 1211.4838)
[20] Álvarez V *et al.* (NEXT Collaboration) 2012 *Nucl. Instrum. Meth. A* (*Preprint* 1211.4474)
[21] Oliveira C A B 2013 *J. Phys. Conf. Ser.* (*Preprint* 1301.XXXX)
[22] Nygren D 2009 *Nucl. Instrum. Meth. A* **603** 337–348
[23] Álvarez V *et al.* (NEXT Collaboration) 2012 *JINST* **7** P02010 (*Preprint* 1201.2018)
[24] Luscher R *et al.* 1998 *Phys. Lett. B* **434** 407–414
[25] Álvarez V *et al.* 2013 *JINST* **8** T01002 (*Preprint* 1211.3961)
[26] Malling D *et al.* 2011 (*Preprint* 1110.0103)
[27] Aprile E *et al.* (XENON Collaboration) 2011 *Astropart. Phys.* **35** 43–49 (*Preprint* 1103.5831)

Journal of Physics: Conference Series **460** (2013) 012011 doi:10.1088/1742-6596/460/1/012011

Results from the NEXT prototypes

C A B Oliveira, on behalf of the NEXT collaboration
Lawrence Berkeley National Laboratory
One Cyclotron Road, Berkeley, CA 94720, USA
E-mail: cabdoliveira@lbl.gov

Abstract. NEXT-100 is an electroluminescent high pressure Time Projection Chamber currently under construction. It will search for the neutrino-less double beta decay in ^{136}Xe at the Canfranc Underground Laboratory. NEXT-100 aims to achieve nearly intrinsic energy resolution and to highly suppress background events by taking advantage of the unique properties of xenon in the gaseous phase as the detection medium. In order to prove the principle of operation and to study which are the best operational conditions, two prototypes were constructed: NEXT-DEMO and NEXT-DBDM. In this paper we present the latest results from both prototypes. We report the improvement in terms of light collection ($\sim 3\times$) achieved by coating the walls of NEXT-DEMO with tetraphenyl butadiene (TPB), the outstanding energy resolution of 1 % (Full Width Half Maximum) from NEXT-DBDM as well as the tracking capabilities of this prototype (2.1 mm RMS error for point-like depositions) achieved by using a square array of 8×8 SiPMs.

1. Introduction

If the neutrino has a Majorana nature, the neutrino less double beta ($0\nu\beta\beta$) decay exists in some natural isotopes including ^{136}Xe. The observation of this hypothetical decay would prove physics beyond the Standard Model as the neutrino being its own anti-particle and the lepton number violation. In addition, the measurement of its lifetime would allow to estimate the absolute neutrino masses and to determine the hierarchy of the neutrino mass states.

NEXT-100 is currently under construction and will search for the $0\nu\beta\beta$ decay at the Canfranc Underground Laboratory [1]. It consists of an electroluminescent high pressure (up to 15 bar) Time Projection Chamber that will use up to 150 kg of gaseous xenon enriched in the isotope ^{136}Xe to ~ 91 % as both the source of the decay and the detection medium.

In the gaseous phase the fluctuations associated with the production of primary electrons by ionizing particles are very low when compared to those of the liquid phase ($\sim$130 times higher) [2]. Taking advantage of this and using electroluminescence (EL) as a nearly noiseless signal amplification process [3] allows one to achieve outstanding energy resolution, which plays an important role in separating the rare signal from natural background decays that are close in energy. In addition, the extended ionization tracks left behind by the two high energy electrons from the decay ($\sim$ 30 cm long at 10 bar) enable the suppress of background by performing topology recognition of the "spaghetti with two meat-balls" signature [4]. For further details about NEXT-100, its working principle and its physics potential please see references [5]and [6].

As part of the R&D program of the NEXT experiment, two prototypes were constructed: NEXT-DEMO and NEXT-DBDM. Both prototypes have a similar working principle (and similar to that of NEXT-100) but the studies performed in each one focused on different issues. Unlike

NEXT-100, the two prototypes are not intended to be radiopure neither shielded from natural radioactivity. In this paper we summarize the latest results from both prototypes regarding gas purity stability, light collection, energy resolution and tracking capabilities. Detailed analyses are available in references [7] and [8] for NEXT-DEMO and NEXT-DBDM, respectively.

2. Principle of operation and data acquisition

When incoming particles interact inside the detector, they ionize and excite atoms of the gas. Excitations produce a prompt and fast light (172 nm) pulse – called S_1 – that, after being detected by an array of photomultipliers (PMTs), works as a *start-of-event* signal. The free electrons are then under effect of a drift field that helps them to escape recombination while they drift towards a region – the EL gap – where they are accelerated by a stronger electric field and excite but not ionize xenon atoms, which then de-excite producing secondary light (also 172 nm). This process is called *electroluminescence* (EL) and, being detected by the array of PMTs, this light gives information about the energy of the incident particle – the S_2 signal. EL is also detected by a 2D array of light sensitive detectors (PMTs in the case of NEXT-DEMO and SiPMs in the case of NEXT-DBDM) that lives just next to the EL gap. By time slicing the waveforms coming from those devices, it is possible to image in 3D the original ionization track. As an example of the acquired raw data, Fig. 1 shows the so-called *summed waveform* which consists of the sum of the waveforms recorded by the 19 PMTs of NEXT-DBDM for a given event (662 keV Gamma-ray deposition).

The working principle of the prototypes is very similar to that of the final NEXT-100 detector [5]. For further details about the EL process please see reference [3].

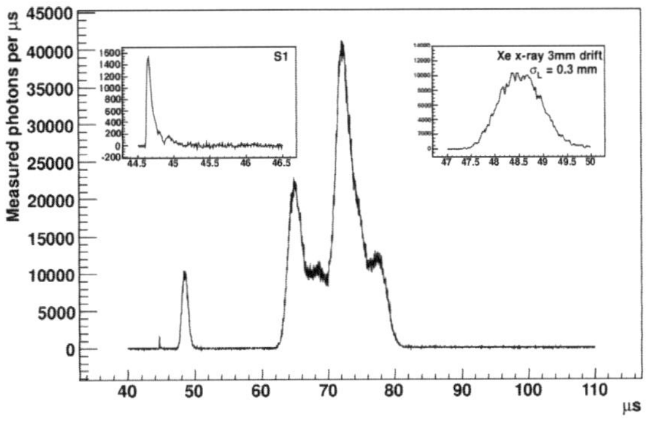

Figure 1. Sum of the waveforms recorded by the 19 PMTs of NEXT-DBDM for a 662 keV Gamma-ray interaction. At $\sim$ 45 μs it is visible the prompt S_1 pulse, at $\sim$ 48 μs a 30 keV xenon X-ray deposition pulse, for $t > 65\mu$s the main event deposition which structure may be affected by the topology of the track, the formation of δ-rays, Compton depositions and multiple scattering dependence on the electron energy.

3. NEXT-DEMO

NEXT-DEMO was constructed and operated at the Instituto de Física Corpuscular (Valencia, Spain). The energy plane consists of an array of 19 R7378A PMTs from Hamamatsu with 1" quartz windows, resistant to pressure up to 20 bar and with a quantum efficiency (QE) of about 20 %. It is placed 10 cm behind the cathode. A similar array was installed 2 mm away from the anode for coarse tracking. The drift region is 30 cm long and has an hexagonal cross-section with a 14.4 cm long internal diagonal. The length to diameter ratio of the EL light detected by the energy plane is $l/d = 40.0$ cm$/14.4$ cm $= 2.8$. This ratio does not optimise light collection but allowed for the construction of a longer pressure vessel with flexibility for possible upgrades. A 1 μCi ^{22}Na calibration source has been used to characterise the detector. ^{22}Na decays emitting a positron that travels less than 1 mm before annihilating with an electron and emitting two back-to-back 511 keV Gamma-rays. A NaI detector was used as a trigger in order to optimize the fraction of useful data. The setup is in such a way that a large fraction of the Gammas enter the chamber perpendicular to the longitudinal direction, close to the cathode. The prototype scheme is shown in Fig. 2(a) and a picture of the system and the radioactive source + NaI detector setup is shown in Fig. 2(b).

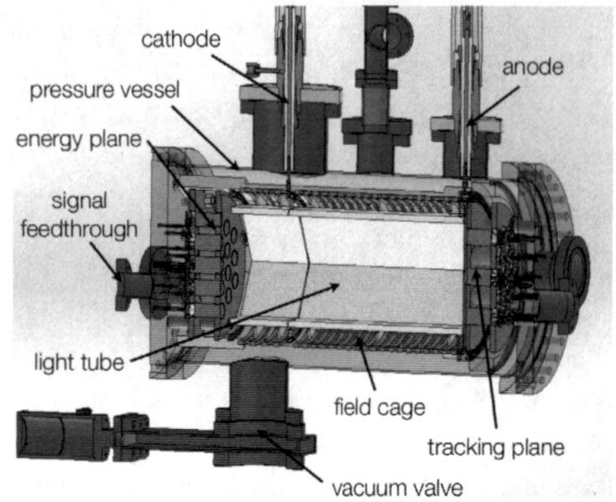

(a) Schematic of the NEXT-DEMO prototype showing the pressure vessel, the energy and coarse tracking PMT planes, the field cage and the feedthroughs.

(b) Picture of the system showing the geometry of the ^{22} Na source and the NaI tagger setup relative to the pressure vessel.

Figure 2. NEXT-DEMO prototype setup.

Two different sets of data were taken with this prototype: the first with the Teflon walls uncoated – the *UV configuration* – and a second one with those walls coated with tetraphenyl butadiene (TPB). TPB shifts 172 nm S_1 and S_2 light to the blue range ($\sim$ 430 nm where the PMTs have higher QE), thus we call this setup the *Blue configuration.* Xenon was recirculated at 10 bar and room temperature through a hot getter (SASE PS4-MT15) to remove impurities. High gas purity was achieved, as confirmed by the smallness of the required corrections for attachment to the measured energy, in the range 1 - 3 % for the whole 30 cm long drift region, depending on whether the walls of the detector were coated or uncoated. A drift field of $\sim$ 667 Vcm^{-1} and an EL field of 2.4 kVcm^{-1}bar^{-1} were used for both sets of data.

3.1. Light collection improvement by TPB

Figure 3 shows, in the left, the spectrum obtained with the UV configuration, after corrections for attachment as well as for radial and angular position dependencies, (r, ϕ), at the level of 0.5 - 3.0 %. Only events with a reconstructed radial position less than 35 mm, where the position response of the detector is linear, are considered. From the left to the right of the x-axis, expressed in number of detected VUV photons - *pes*: at $\sim$ 250 *pes* the 30 keV xenon fluorescence X-ray peak, between 500 and 2700 *pes* the Compton shoulder, for $\sim$ 2800 *pes* the escape peak and for $\sim$ 3000 *pes* the full energy deposition peak. The corresponding spectrum obtained with the Blue configuration is shown in the right of Fig. 3 and shows equivalent features, despite the lower statistics. The number of detected photons increases by a factor of $\sim 3\times$ showing the advantage of coating the detector walls with TPB. The energy resolution of the full energy deposition peak (511 keV) is 2.8 % FWHM for the UV configuration, in agreement with Monte Carlo simulations where only geometric effects were considered. For the Blue configuration,

Journal of Physics: Conference Series **460** (2013) 012011 doi:10.1088/1742-6596/460/1/012011

1.7 % FWHM is achieved, in agreement with the expected scaling due to the increase in the photon statistics. This shows that after calibration, the contributions to the energy response of the detector are well understood.

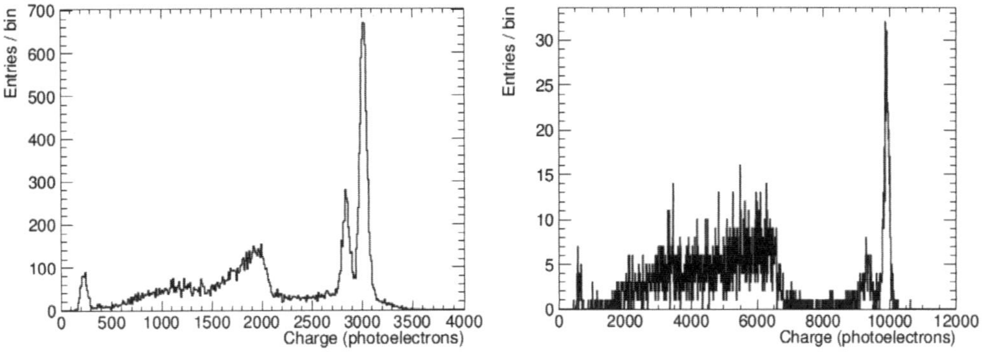

Figure 3. Spectra obtained in NEXT-DEMO with the UV (left) and Blue (right) configurations, after applying corrections for attachment as well as for radial and angular position dependencies.

4. NEXT-DBDM

NEXT-DBDM is a prototype that aims to go beyond the NEXT R&D and explore the possibility of Double Beta and Dark Matter (DBDM) searches using the same detector. It was constructed and is being operated in the Lawrence Berkeley National Laboratory (Berkeley, California, USA). Figure 4(a) shows the schematics of the detector; a picture of the pressure vessel, feed-throughs and vacuum system is shown in Fig. 4(b).

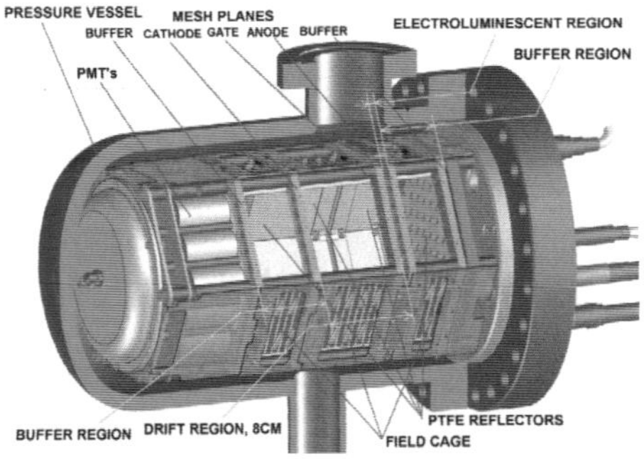

(a) Schematic of the NEXT-DBDM prototype showing the pressure vessel, the energy plane, the field cage and the cathode and EL meshes.

(b) Picture of the system showing the pressure vessel, the feed-throughs that service the detector and the vaccum system.

Figure 4. NEXT-DBDM prototype.

The energy plane is made of 19 PMTs (same as those used in NEXT-DEMO) placed 5 cm behind the cathode. The drift region is 8 cm long with an hexagonal cross section of 14.4 cm diameter. The length to diameter associated with the EL light is l/d = 13.0/14.4cm = 0.9, optimizing light collection. This allowed us to perform detailed energy response studies envisaging nearly intrinsic resolutions. The results reported in this document refer to events produced by a 1 mCi ^{137}Cs 662 keV Gamma-ray source, after being strongly collimated and directed along the axis of the chamber.

Journal of Physics: Conference Series **460** (2013) 012011 doi:10.1088/1742-6596/460/1/012011

Xenon at room temperate and pressures of 10 and 15 bar was used, recirculating through a hot getter from Johnson Matthey (model PureGuard) for purification. High quality pure xenon was achieved, indicated by the long electronic life times: $\tau = 36 \pm 10$ ms for 10 bar and $\tau = 9.1 \pm 0.9$ ms for 15 bar. This means that less than 1 % of the primary electrons were lost through attachment along the whole 8 cm long drift region. Drift fields in the range of 10 - 120 $\mathrm{Vcm^{-1}bar^{-1}}$ were scanned without significant changes in the obtained energy resolution. EL fields between 1.0 and 2.5 $\mathrm{kVcm^{-1}bar^{-1}}$ were also scanned with good consistency of the amount of EL light for both pressures.

4.1. Energy studies

Results reported in this Section correspond to runs taken without a dedicated tracking plane. However, a rough (x, y) reconstructed position of the events was possible using the energy PMT plane, with resolutions of ~ 2 mm for central 30 keV X-ray depositions.

Figure 5(a) shows the acquired energy spectrum at 10 atm around the 662 keV energy window, for events reconstructed in the central 0.6 cm radius region. The energy response was corrected only for the small attachment losses. The xenon X-ray escape peak is clearly visible 30 keV below the full energy deposition peak. At 15 atm, a world record of 1 % FWHM energy resolution for 662 keV Gamma-rays was obtained. If scaled according to a $\frac{1}{\sqrt{E}}$ dependency, this resolution extrapolates to 0.52 % at $Q_{\beta\beta}$, well below the initial target of the NEXT experiment (1 %).

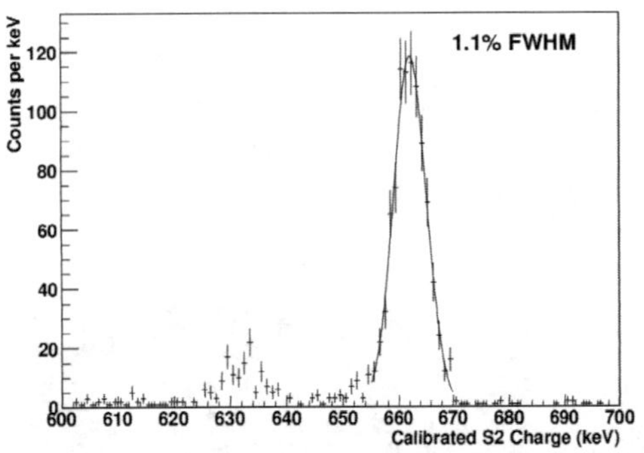

(a) Energy resolution for 662 keV depositions at 10.1 atm. The data was acquired with a drift and EL fields of 0.16 $\mathrm{kVcm^{-1}}$ and 2.08 $\mathrm{kVcm^{-1}atm^{-1}}$, respectively. The events were corrected for small attachment losses ($\tau = 13.9$ ms).

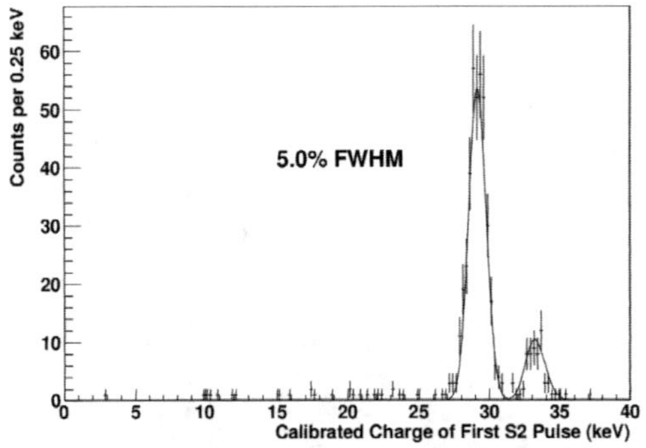

(b) Energy spectrum for fluorescence xenon X-rays acquired at 10.1 bar. The data was acquired using a drift field of 1.03 $\mathrm{kVcm^{-1}}$ and an EL field of 2.68 $\mathrm{kVcm^{-1}atm^{-1}}$.

Figure 5. Energy resolution with NEXT-DBDM.

Figure 5(b) shows the spectrum for fluorescence X-rays pulses that were well separated in time from the main deposition and were reconstructed in the central 1.5 cm radius region. For the K_α peak, with a nominal energy of 29.6 keV [9], an energy resolution of 5 % FWHM was achieved. The centroid of this peak, after energy calibration on the 662 keV full energy peak (and considering linearity with a zero intercept), is at 29.1 keV, less 2 % away from the nominal value. The energy resolution at these energies is good enough to clearly resolve separately the K_α and K_β (nominally at 33.6 keV) lines of X-ray shell transitions in xenon.

The plot of Fig. 6 summarises our knowledge about the energy response in the NEXT-DBDM prototype. The energy resolution dependence on the amount of detected light in an electroluminescent TPC is described by four contributions. The first is the gas intrinsic energy resolution given by the Fano factor [10] and the energy of the incoming particle, $F = 0.14$ for our conditions. It is represented in the plot by the two horizontal lines, for 30 keV X-rays and 662 keV Gamma-rays respectively. The process of EL amplification is a nearly noiseless process and thus does not contribute significantly to the energy resolution degradation [3]. The conversion of photons into photoelectrons in the PMTs photocathode follow a Poisson distribution represented by the lower diagonal line. The upper diagonal line shows the expected response when including the variance in the PMT response to single photoelectrons, which we determined previously. The data points taken with a pulsed LED (circles) follow that trend.

Summing all the contributions one gets the two curved lines, respectively for 30 and 662 keV depositions. The data points follow the expected tendency but are larger than the expected form (20-30 % in the case of 662 keV Gammas) possibly due to the (x, y) response non-uniformity. It is expected that a dense tracking array as that recently installed in the prototype, which preliminary results are presented in Section 4.2, will allow one to map that non-uniformity and correct the detector response, further enhancing the energy resolution.

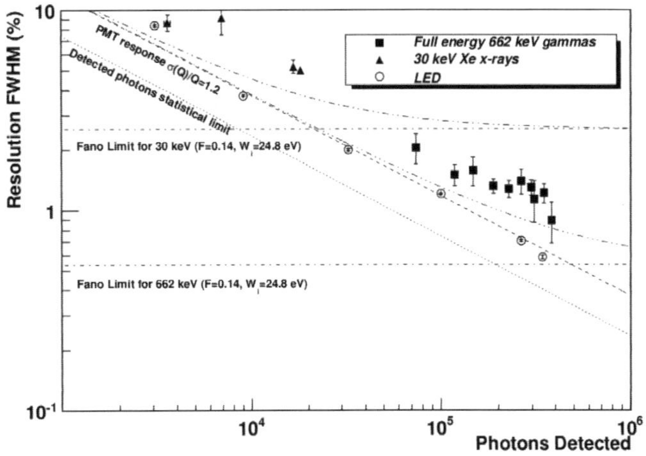

Figure 6. Understanding the energy response of the NEXT-DBDM: points show the measured energy resolutions for 662 keV Gammas, 30 keV xenon X-rays and LED light pulses as a function of the number of detected photons. The expected resolution is shown for X-rays (dot dot dashed) and for 662 keV Gammas (dot dot dot dashed). The results from 662 keV depositions were obtained from 15.1 atm runs while fluorescence X-ray results were obtained from 10.1 atm runs.

4.2. Tracking

In this Section, we report tracking results achieved after installing 64 Hamamatsu S10362-11-025P SiPMs, placed 2 mm behind the anode. These devices have 1 mm^2 active area and were arranged in a 8 × 8 square array with a pitch of 1 cm, as shown in Fig. 7(a). The SiPMs have been coated with TPB so they can be sensitive to the 172 nm light from xenon EL [11]. A set of 4 Dice-boards provide support for the SiPMS as well as connections for high voltage supply and signal readout. Figure 7(b) shows the low power electronics used for signal amplification, shaping and multiplexing of the SiPMs waveforms. These waveforms (not calibrated for the results presented here) are sliced in 30 keV energy equivalent sections and used to perform likelihood fits in order to determine the position of such partial depositions.

Figure 8(a) shows the relative S_2 intensities detected in each SiPM for an externally tagged cosmic ray muon traversing the chamber. The squares are the reconstructed (x, y) positions of each section of the sliced data and show the expected straight muon trajectory. The estimated position RMS error is estimated to be 2.1 mm. In Fig. 8(b), a similar plot is shown but for the case of a 662 keV Gamma-ray event. The main deposition in the center of the chamber is clearly

(a) Picture of the 8 × 8 array of SiPMs supported by 4 Dice-boards, placed behind the anode mesh. For a scale reference, the inner diagonal of the hexagonal frame is 14.4 cm long.

(b) Low power electronics SiPM readout. A front-end card (at the foreground), used for signal amplification and shaping, connects to 4 SiPMs. Four of these connect to a multiplexing board (in the left).

Figure 7. Tracking system of NEXT-DBDM.

visible. It is about 3 cm long and affected by the expected multiple scattering that makes the track curled. Around the position (4, 1.5) cm, it is also visible a well separated X-ray, about 5 cm away from the main deposition. These results, which can still be improved, were obtained without calibration and with no optimized analysis, but show that tracking with SiPMs works in a gaseous electroluminescent TPC, enabling topological identification of the $0\nu\beta\beta$ decay.

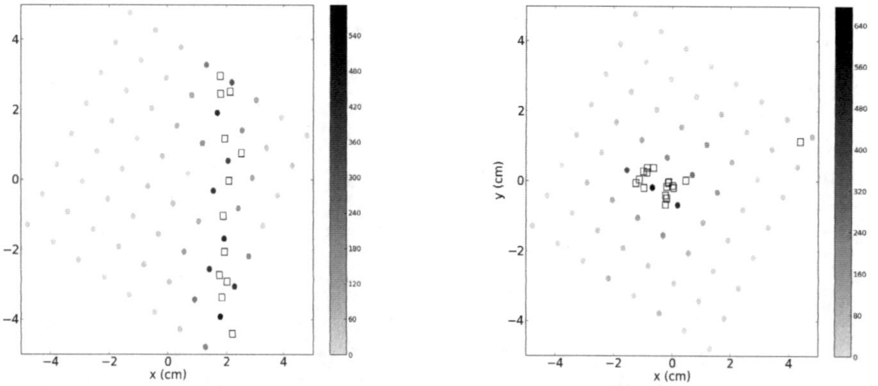

(a) 2D image of the track of an externally triggered muon. Circular points represent the S_2 intensities of each SiPM, according to the gray scale. Square points represent the reconstructed (x, y) positions of 30 keV point-like depositions.

(b) 2D image of the deposition of a Gamma-ray. The main deposition at the center of chamber is visible along with a well separated fluorescence X-ray deposited in the up right edge of the SiPM array.

Figure 8. Tracking with NEXT-DBDM.

5. Other developments

The drift velocity and longitudinal diffusion of electrons in pure xenon has been assessed in NEXT-DEMO using alpha particles, which produce point-like depositions and show a strong anti-correlation between the ionization and the scintillation signals. Details about this work may be found in reference [12]. Using NEXT-DBDM and 30 keV fluorescence X-rays (also point-like) we measured the electron drift parameters in mixtures of xenon and CF_4, achieving good consistency with NEXT-DEMO and Monte Carlo results. Measurements of Xe+CH_4 are also planned and will be reported in a future publication.

Another mixture explored by the collaboration is Xe+TMA (tri-methyl-amine). Such a mixture (in a 98.5/1.5 proportion) has been recently employed with advantage for the operation of Micromegas amplification structures. A behavior markedly suggestive of the presence of Penning-like transfer processes was found [13] which may be related with the observed improved energy resolution (7.3 % FWHM for 22 keV at 1 bar and 9.6 % FWHM at 10 bar) [14]. The collaboration is currently employing this mixture in a medium-size (30 cm-diameter readout / 38 cm-length drift) cylindrical TPC (NEXT-MM) aiming at the characterization of the electron swarm properties, as well as exploring the ultimate limits of Micromegas operation at high pressure. Raw full-absorption energy peaks and the corresponding curled electron tracks stemming from various X-ray sources have been presented at the conference. A detailed analysis (including the necessary calibrations and corrections) will be presented elsewhere.

6. Conclusions

In the last 3 years the NEXT collaboration carried out a R&D program with the aim of demonstrating the NEXT detector concept and gaining experience that would facilitate the design, construction and operation of the NEXT-100 detector. Two prototypes, NEXT-DEMO and NEXT-DBDM, were constructed and operated, with complementary goals.

NEXT-DEMO has shown the advantages of coating the TPC walls with TPB with a 3.0× improvement in S_1 and S_2 light collection. Outstanding energy resolution of 1.0 % FWHM for 662 keV Gamma-rays was achieved with NEXT-DBDM for the range of xenon pressures that will be used in NEXT-100. This energy resolution extrapolates to 0.52 % at $Q_{\beta\beta}$, well below the initial 1 % goal of NEXT. Tracking using SiPMs for measuring the EL signal works and will allow further improvements in the energy resolution after correction of the S_2 position dependencies.

These promising results represent a milestone towards the success of NEXT-100 and prove the working concept of such a detector.

Acknowledgments

This work was supported by the Director, Office of Science, Office of Basic Energy Sciences, of the U.S. Department of Energy under Contract No. DE-AC02-05CH11231. This work used resources of the National Energy Research Scientific Computing Center (NERSC). This work was also supported by the Ministerio de Economía y Competitividad of Spain under grants CONSOLIDER-Ingenio 2010 CSD2008-0037 (CUP) and FPA2009-13697-C04-04.

References

[1] *LSC - Latoratorio Subterraneo de Canfranc*, http://www.lsc-canfranc.es/en/
[2] A Bolotnikov and B Ramsey, 1997, *Nucl. Instr. and Meth. A* **396** 360
[3] C A B Oliveira et al, 2011, *Physics Letters B* **703** 217
[4] The NEXT Collaboration, 2011, *arXiv:1106.3630v1 [physics.ins-det]*
[5] The NEXT Collaboration, 2012, *JINST* **7** T06001
[6] J Martín-Albo and J J Gómez-Cadenas, *J. Phys. Conf. Ser.* (Preprint *arXiv:1301.2966v1 [physics.ins-det]*)
[7] The NEXT Collaboration, 2012, *arXiv:1211.4838v2 [physics.ins-det]*
[8] The NEXT Collaboration, 2013, *Nucl. Instr. and Meth. A*, **708** 101.
[9] L P Ekstrom, R B Firestone, http://ie.lbl.gov/toi/xraySearch.asp
[10] U Fano, 1947, *Physical Review*, **72** 26.
[11] The NEXT Collaboration, 2012, *JINST* **7** P02010
[12] The NEXT Collaboration, 2013, *arXiv:1211.4508v1 [physics.ins-det]*
[13] S Cebrián et al., 2013, *JINST* **8** P01012.
[14] D Nygren, 2011, *J. Phys: Conf. Ser.* **309** 012006.

Journal of Physics: Conference Series **460** (2013) 012012 doi:10.1088/1742-6596/460/1/012012

Micromegas-TPC operation at high pressure in Xenon-trimethylamine mixtures

D. C. Herrera[a], S. Cebrián[a], T. Dafni[a], E. Ferrer-Ribas[b], I. Giomataris[b], D. Gonzalez-Diaz[a,c], H. Gómez [a,d], F.J. Iguaz[a], I.G. Irastorza[a], G. Luzon[a], A. Rodríguez[a], L. Segui[a], A. Tomás[a]

[a] Laboratorio de Física Nuclear y Astropartículas, Universidad de Zaragoza, Zaragoza, Spain
[b] IRFU,Centre d'Etudes Nucleaires de Saclay (CEA-Saclay) 91191 Gif sur Yvette, France
[c] Department of Engineering Physics, Tsinghua University, Liuqing building, Beijing, China
[d] Laboratoire de l'Accélérateur Linéaire, Université Paris-Sud 11, Orsay Cedex (FRANCE)

E-mail: `diana.he@unizar.es`

Abstract. We present in this work measurements performed with a small Micromegas-TPC using a xenon-trimethylamine (Xe-TMA) Penning-mixture as filling gas. Measurements of gas gain and energy resolutions for 22.1 keV X-rays are presented, spanning several TMA concentrations and pressures between 1 and 10 bar. Across this pressure range, the best energy resolution and largest increase in gain at constant field (a standard figure for characterizing Penning-like energy transfers) is observed within the 0.9%-1.7% TMA range. A gain increase (at constant field) up to a factor 100 and best values of the energy resolution improved by up to a factor 3 with respect to the one previously reported in pure Xe -operated Micromegas, can be obtained. In virtue of the VUV-quenching properties of the mixture, the overall maximum gain achievable is also notably increased (up to 400 at 10bar), a factor ×3 higher than in pure Xe. In addition, preliminary measurements of the electron drift velocity in a modified setup have been performed and show good agreement with the one obtained from Magboltz.

These results are of great interest for calorimetric applications in gas Xe TPCs, in particular for the search of the neutrino-less double beta decay ($0\nu\beta\beta$) of ^{136}Xe.

It is important to note that in this work some figures from [1] have been updated. Precisely, the TMA concentration has been re-estimated after a detailed re-calibration of the mass spectrometer yielding lower TMA concentration values, corrected by a factor in the range $0.5-0.7$. An erratum to [1] is being submitted at the moment of writing.

1. Introduction

The detection of the $0\nu\beta\beta$ decay in ^{136}Xe can provide both the neutrino mass scale and an unambiguous answer to its nature (Majorana or Dirac). For this reason, the development of large gas Xe Time Projection Chambers (TPCs) targeted at finding evidence for $0\nu\beta\beta$ in ^{136}Xe is currently active. This kind of detector can fulfill the main requirements of the current generation of experiments (100 kg Xe), which are an energy resolution down to 1% FWHM (full width half maximum) at the excess energy of the decay ($Q_{\beta\beta}$), and topological information from track reconstruction. Moreover, it can be easily scaled up to 1 ton, thus allowing for the exploration of the entire inverse hierarchy of neutrino mass models. The NEXT experiment (Neutrino Experiment with Xenon TPC) will search the $0\nu\beta\beta$ decay of ^{136}Xe at the Canfranc Underground Laboratory using 100 kg of enriched Xe in an electroluminescent high pressure

TPC [2, 3]. As part of the R&D program of the NEXT experiment, various prototypes were built, one of them based on charge collection from ionization using Micromegas detectors [4, 5]. Specifically, the microbulk technique [6] has been used in several experiments due to its excellent position resolution and good energy resolution, as well as robustness and low radioactivity [7, 8], being this latter feature more important for rare events searches [9].

We focused on the possibility of improving the energy resolution of the Micromegas-TPC using Xe-TMA Penning mixtures. Early studies performed in wire chambers at 1 bar [10] showed indeed a great gain enhancement and improved energy resolution. We performed an experimental study of this mixture which, it is believed, can also provide a starting point for evaluating the recent proposal of using TMA as additive gas to reduce the Fano factor in Xe-based TPCs [11]. All results conveyed in this work have been obtained with a small-size TPC, that is later described. A medium size 73 l TPC (NEXT-MM) has been built and commissioned based on the microbulk technology, and its preliminary tracking performances presented at the conference. This topic will be the subject of a dedicated paper.

The main characteristics of the experimental setup are described in section 2, while in section 3 and 4 the experimental procedure and measurements are summarized. Finally, in section 5 the conclusions and outlook is presented. A more detailed description of these measurements can be found in [1].

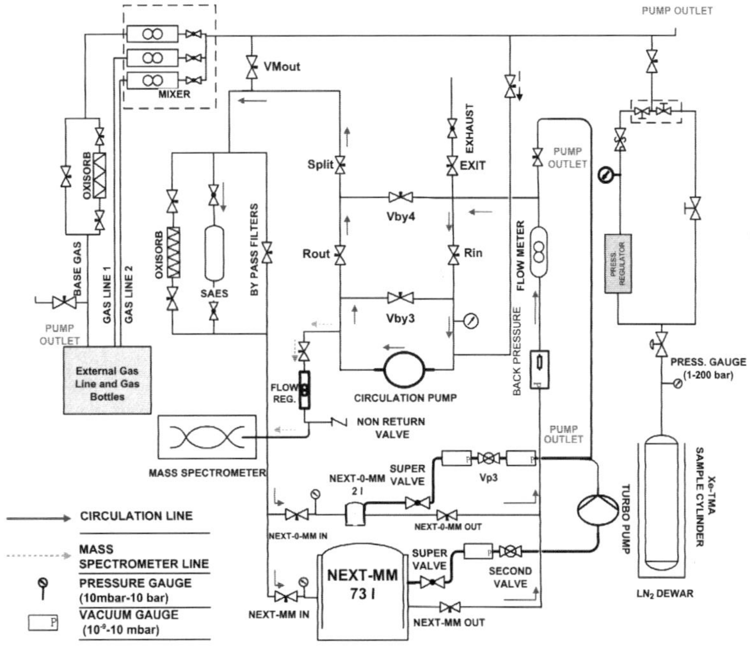

Figure 1. Schematic view of the experimental setup, underlining the gas system. At the bottom/center, the TPC employed for these measurements (dubbed 'NEXT-0') is shown. A medium-size TPC (NEXT-MM), central to this system, will be subject of a forthcoming communication.

2. Experimental setup

The experimental setup consists of a small TPC and a mass spectrometer -monitored gas system that allows, amongst other features, for high-pressure recirculation, gas filtering and cryo-pumping. A schematic view of the experimental setup, mainly underlining the gas system, is shown in Fig. 1, where the small TPC has been dubbed 'NEXT-0', described in more detail in [7]. The TPCs inserted in the gas system are built of stainless steel with materials of low outgassing and have been tested up to 12 bar. The small TPC has an inner volume of 2.4 l (10 cm height, 16 cm diameter). The drift distance, equipped with copper rings, can be varied between 1 and 6 cm.

For these measurements, a special purifier (SAES 702) was installed in order to work with TMA, therefore water vapor and electronegative impurities (H_2O, O_2, CO_2) were constantly removed during the recirculation process. The gas recovery system has a stainless steel sample cylinder of 2.2 l that is immersed in a Dewar flask filled with liquid nitrogen (LN_2) during the recovery process. A Pfeiffer OmniStar mass spectrometer properly calibrated for different Xe+TMA concentrations is used to quantify the gas composition of Xe+TMA mixtures and to monitor the electronegative impurities.

3. Experimental procedure and results

The first experimental goal was to establish a range of TMA concentration for which it would be possible to obtain the best energy resolution and the highest gain for pressures between 1 and 10 bar. Hence, a systematic variation of the TMA concentration was performed at four reference pressures: 1, 5, 8 and 10 bar (see section 3.2). Once the optimal TMA concentration range was established, a systematic study at various pressures in a range from 1 to 10 bar was realized (see section 3.3).

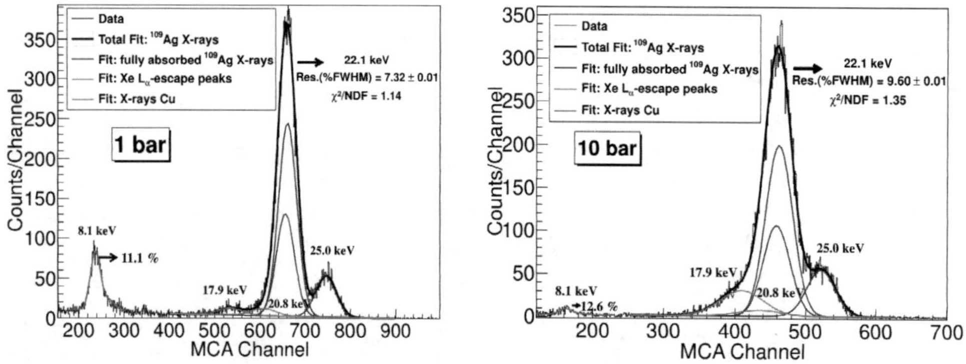

Figure 2. X-ray energy spectra generated by a ^{109}Cd source, in a Xe+1.2 % TMA mixture at 1 bar (left) and 0.7% TMA at 10 bar (right). The fit performed to the overall ^{109}Ag K-fluorescence lines consists in a 3-step routine. In addition, the Cu K-fluorescence at 8.1 keV is separately fitted to a single Gaussian function.

3.1. Experimental procedure

All measurements were carried out with a ^{109}Cd collimated source placed at the center of the cathode plate. Prior to the experimental campaign, a highly concentrated Xe+TMA mixture (93/7) was prepared in the sample cylinder. For all subsequent measurements the mixture was constantly passed through a purifier, and its exact content adjusted by injection of fresh gas whenever needed. Contrary to the specifications of the provider, it was observed that the SAES filter absorbs or expels TMA, in direct relation to the history of the previous concentrations employed. Hence it was necessary to wait approximately 30 minutes until the mixture was homogenized. Once the gas mixture was stabilized, the operating point was established. For gas gain measurements the amplification field was systematically increased until two consecutive sparks were observed to occur within a short time (30 s). At the end of each set of measurements, the TMA concentration was determined with the mass spectrometer. The chamber and the gas system were cleaned by recovering the gas and then the complete system was pumped.

In the off-line analysis, the ^{109}Ag X-ray peak at 22.1 keV was used to obtain the gain and the energy resolution. In Fig. 2, the typical energy spectrum acquired with a ^{109}Cd source is shown at 1 bar in a Xe+1.2% TMA mixture (left) and at 10 bar with 0.7% TMA (right). In both spectra the K_α and K_β lines from the Ag fluorescence are clearly distinguished. The corresponding escape peaks from Xe are observed below the K_α line, located at 17.9 keV and 20.8 keV. The Ag K_α, K_β lines (in blue) and the corresponding Xe escape peaks from the L-shell (in magenta) were fitted in the energy range between 14 and 30 keV using a 3-step routine (see Fig. 2). On the other hand, it is also observed the Cu K-fluorescence at 8.1 keV, which is produced from the interaction of X-rays with the electrodes of the Micromegas.

Fig. 3 shows an energy spectrum acquired at 8 bar using a mixture of Xe+0.8% TMA, with a larger energy range than the previous figure. It is interesting to observe the γ-rays from the ^{109}Cd source at 88.04 keV (green) and the two associated Xe escape peaks in magenta (with 'escaping' energy totalling that of the K_α and K_β Xe-lines). As expected, the energy resolution is roughly inversely proportional to $\sqrt{E}$ such as shown in Fig. 3 (right).

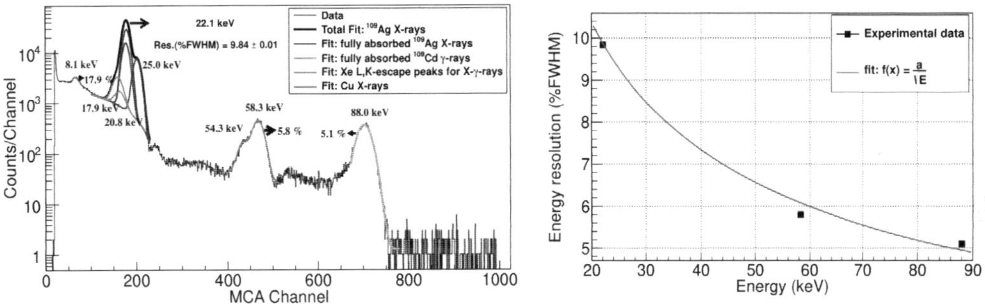

Figure 3. Left: energy spectrum from a ^{109}Cd source acquired at 8 bar in a Xe+0.8% TMA mixture, with $E/p =$ 245 V/cm/bar in the drift region. The K-fluorescence emission and the γ-rays from ^{109}Ag with their corresponding escape peaks of Xe are shown. Right: behaviour of the energy resolution for the photon energies emitted by a ^{109}Cd source, showing an approximate $1/\sqrt{E}$ scaling.

3.2. Optimum concentration of TMA

The variation in TMA concentration was performed in different ranges: at 1 bar (0.3 % – 10.4 %), 5 bar (0.3 % – 3.8 %), 8 bar (0.2 % – 3.3 %), and 10 bar (0.4 % – 3.8 %). The dependence of the gain with the amplification field is shown in Fig. 4 for two reference pressures: 1 (a) and 10 (b) bar. The gain curves show a linear behaviour with the amplification field in the semi-log plot, a fact often interpreted as indicative of negligible feedback. Considering the plot at 1 bar (see Fig. 4a), we observe that lower amplification fields must be applied when the TMA percentage is increased from 0.3% to 0.9%. The curves within a concentration range from 0.9% and 4.5% TMA seem to overlap, suggesting that transfer mechanisms are already fully active while the avalanche dynamics remains largely unaffected. The tendency changes above 4.5% TMA so

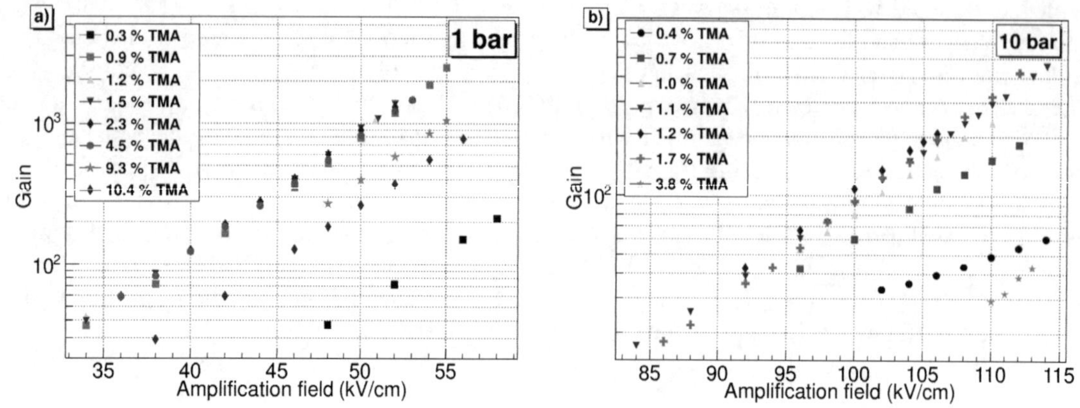

Figure 4. Dependency of the gas gain on amplification field for different TMA concentrations at 1 (a) and 10 (b) bar.[1]

that higher fields must be applied to obtain the same gas gain, as also observed in neon-based mixtures [12]. A similar behaviour is seen at high pressures, exemplified here for the systematic measurements at 10 bar (Fig. 4b). The increased energy loss by inelastic collisions to TMA molecules is the most likely explanation for this change in tendency, therefore shattering the extra ionization obtained by Penning transfer.

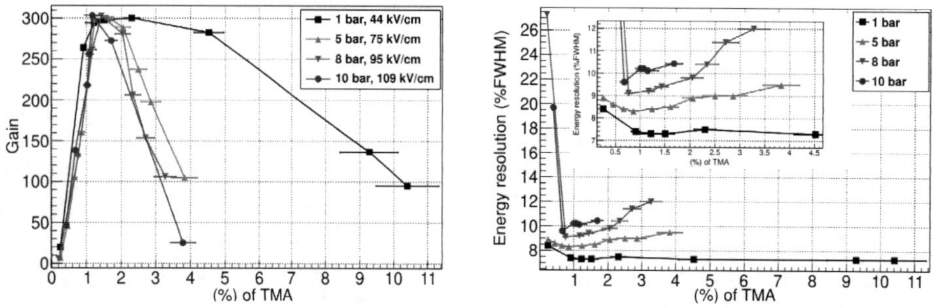

Figure 5. Gas gain at a fixed amplification field (left) and best energy resolution for each mixture as a function of TMA concentration. For each pressure, an optimum region exists where the highest values in gas gain and best energy resolutions are simultaneously achieved.[1]

In order to do a better study of the optimum TMA concentration, we performed linear fits of the $\ln G$ versus amplification field data for a proper interpolation or extrapolation of the gain. The variation of the gas gain (at a constant amplification field) with the percentage of TMA is shown in Fig. 5 (left). At each pressure the gas gain rapidly rises when small quantities of TMA are added up to values around 1.2% TMA. Further, a range where the gain remains roughly constant occurs between 1.2% and 1.8% TMA, except for 1 bar where the gain remains constant in a larger range ($\sim$ 0.9%-3.0%). The gas gain decreases sharply thereafter. At low pressures

[1] Figures have been updated from [1]. The TMA concentrations were re-estimated after a detailed re-calibration of the mass spectrometer, yielding TMA concentration values lower than the ones presented in the previous work [1], by a factor $\times 0.5 - 0.7$.

(up to 1 bar), the rapid rise in gas gain at constant amplification field has also been observed using Xe+2, 3 dimethyl-2-butene as Penning additive, showing similar dependencies with the additive at 1 bar [13]. This great increase is an evidence that Penning effect takes place.

Figures showing the energy resolution as a function of amplification field are not presented in this work (but can be found in [1]). However, the best energy resolution for each mixture is shown in Fig. 5 (right) as a function of the TMA concentration. In general, we can see that the energy resolution improves with the TMA concentration, the best values of energy resolution being found between 0.8% and 1.6% for each pressure. Therefore, we can conclude that the optimal TMA percentage ranges are found between 0.9% 1.7%, where the best energy resolutions and the highest gains (at constant field) are obtained.

3.3. Varying the pressure

In this section we present the results of the gas gain and the energy resolution obtained when the pressure is varied from 1 to 10 bar, using TMA concentrations within the optimal range that was estimated in the previous section (from 0.9 % to 1.7 % TMA). Gas gain curves are shown in Fig. 6 (left). It is observed that the maximum gain drops nearly exponentially for pressures above 2 bar, down to $\sim$ 400 at 10 bar. However, the maximum gain at any pressure is still at least a factor 3 higher than for Micromegas operated in pure Xe [14].

The results of this work are compared in Fig. 6 (right) with previous measurements with Micromegas detectors in pure Xe: with the same setup used in this study (▲) [16] and with a different one (▼) [14]. The energy resolution achieved at 22.1 keV is substantially better in this work, going down to 7.3 % (9.6 %) FWHM at 1 (10) bar. This fact translates into an improvement of a factor 2 (3) at 1 (10) bar as compared to previous measurements in pure Xe. We can infer that the addition of TMA to Xe reduces the avalanche fluctuations mainly due to the efficient Penning transfer between Xe excited states and TMA molecules (causing ionization of the latter) together with the efficient reduction of photon feedback.

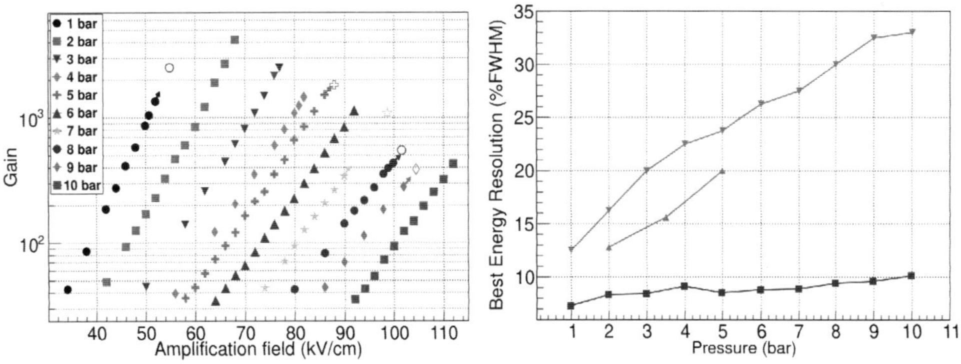

Figure 6. Left: dependence of the gain with the amplification field. The empty marker represents the maximum gain reached for each pressure, in case it was not determined for the given conditions but for a slightly different gas mixture around the optimum. Right: dependence of the best energy resolution with pressure at 22.1 keV, for data from [14] (▼), from [16] (▲) and from this work (■).

4. Coincidence setup

The small-TPC setup was modified to measure the drift velocity and attachment effects in Xe+TMA mixtures. A schematic view of the setup used for these measurements is shown in

Fig. 7 (left). A silicon photo-diode detector together with an ^{241}Am source was encapsulated into a plastic piece (made of POM, with low outgassing), and then installed inside the TPC. The ^{241}Am source emits in coincidence an α-particle and a γ-photon which are detected by the silicon diode and the Micromegas detectors, respectively. Both signals are pre-amplified and registered by a Tektronix oscilloscope, that allows to select the coincidences. With this configuration we obtain a coincidence system which provides the t_0 of each event, hence the drift time and the drift velocity can be obtained. The drift time of each event is calculated from the temporal difference between the α and γ signals. The drift velocity is then obtained from the overall range of drift times (ΔT), that spans the full drift region, and the (fixed) drift distance (Δx) as $v_e = \Delta x/\Delta T$. We have in this way determined the drift velocity for various Xe+TMA mixtures and pressures. In Fig. 7 (right) our preliminary measurements (no error bars) are compared with the Magboltz calculations.

On the other hand, the dependence of Micromegas amplitude with the temporal distance between signals can be used to study attachment effects in our gas. In principle, in absence of attachment the signal amplitude should be independent from the drift time, but will show an exponential behaviour otherwise, whose exponent (-t/τ_e) provides the electron lifetime, τ_e. We have preliminary estimated an electron lifetime (1/(attachment coefficient $\times$ drift velocity)) larger than $\tau_e = 4$ ms with this setup (up to 6 bar). Final results will be present elsewhere.

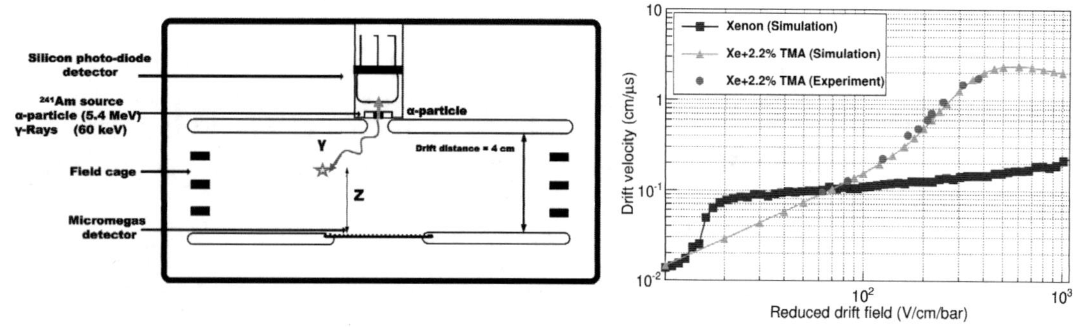

Figure 7. Left: modified setup designed for measuring the drift velocity and attachment. Right: measured drift velocity together with the ones calculated with Magboltz-10 for Xe+TMA mixtures and for pure Xenon. The red markers represent the drift velocity measured for a Xe+2.2% TMA mixture. On the other hand, the green curve represents the drift velocity calculated with Magboltz for Xe+2.2% TMA mixture.

5. Conclusions and outlook

We have performed systematic measurements of gain and energy resolutions in a small TPC filled with Xe-TMA on a large range of concentrations and pressures. We have found that, between 1 and 10 bar, the Penning transfer is optimal for Micromegas operation in the 0.9%-1.7% TMA concentration range, allowing to obtain the best energy resolutions and the highest gain increase at constant field. Energy resolutions down to 7.3% (9.6%) FWHM at 1 (10) bar for 22.1 keV can be achieved, which imply an improvement of a factor 2 (3) with respect to values previously obtained also with microbulk technology, but in pure Xe [14]. This result extrapolates into an energy resolution of 0.7% (0.9%) FWHM at the $Q_{\beta\beta}$ value of Xe for 1 (10) bar, and therefore opens very good prospects for $0\nu\beta\beta$ decay experiments. In addition, we have performed a modification of this setup to measure the drift velocity and attachment. First measurements of the drift velocity in Xe+TMA mixtures have been made and show good agreement with Magboltz calculations.

As part our current work, we have performed measurements in Xe-TMA mixtures in a medium-size TPC (73 l) at 1 bar using microbulk technology. Preliminary results have permitted to observe the first tracks reconstructed of X-rays, γ-emission and background events. The detector performance and the purification system are currently being studied for future improvements. A paper is under preparation to describe in detail the setup, a complete study of the topology of the tracks, energy resolution and gas gain [17].

Acknowledgments

We are grateful to our colleagues of the groups of the University of Zaragoza, CEA/Saclay as well as the NEXT and RD-51 collaborations. We thank R. de Oliveira and his team at CERN for the manufacturing of the microbulk readouts. We acknowledge support from the European Commission under the European Research Council T-REX Starting Grant ref. ERC-2009-StG-240054 of the IDEAS program of the 7th EU Framework Program. We also acknowledge support from the Spanish Ministry of Economics and Competitiveness (MINECO), under contracts ref. FPA2008-03456 and FPA2011-24058, as well as under the CUP project ref. CSD2008-00037 and the CPAN project ref. CSD2007- 00042 from the Consolider-Ingenio 2010 program of the MICINN. Part of these grants are funded by the European Regional Development Fund (ERDF/FEDER). F.I. acknowledges the support from the Eurotalents program.

References

[1] Cebrián S et al. 2013 *JINST* **8** P01012
[2] Alvarez V et al. 2012 *JINST* **7** T06001
[3] Oliveira C A B F et al. [NEXT Collaboration] 2013 arXiv:1302.3902v1 [physics.ins-det]
[4] Dafni T et al. 2011 *J. Phys.: Conf. Ser* **309** 012009
[5] Segui L et al. 2012 *Nucl. Instrum. Meth.* A
[6] Andriamonje S et al. 2010 *JINST* **5** P04007
[7] Cebrián S et al. 2010 *JCAP* **10** 010
[8] Cebrian S et al. 2011 *Astropart. Phys* **34** 354
[9] Irastorza I G et al. 2012 *EAS Publications Series* **53** 147
[10] Ramsey B and Agrawal P 1989 *Nucl. Instrum. Meth.* A **278** 576
[11] Nygren D 2011 *J. Phys.: Conf. Ser.* **309** 012006
[12] Iguaz F J, Ferrer-Ribas E, Giganon A and Giomataris I 2012 *JINST* **7** P04007
[13] White K G Fenton A G and Fenton K B 1987 *Nucl. Instrum. Meth.* A **260** 443
[14] Balan C, Freitas E D C, Papaevangeloub T, Giomataris I, Natal da Luza H, Monteiroa C M B and dos Santos J M F 2011 *JINST* **6** P02006
[15] Sakurai H, Ramsey B D and Weisskopfet M C 1991 *Nucl. Instrum. Meth.* A **307** 504
[16] Dafni T et al. 2011 *J. Phys.: Conf. Ser.* **309** 012009
[17] Cebrián S et al. in preparation

Journal of Physics: Conference Series **460** (2013) 012013 doi:10.1088/1742-6596/460/1/012013

Study of neutrino interactions at the T2K near detector

A. Hillairet
(On behalf of the T2K collaboration)
Dept. of Physics and Astronomy, University of Victoria, 3800 Finnerty Rd, Victoria, BC, V8P 5C2 CANADA

Abstract. The Tokai-To-Kamioka (T2K) experiment was designed to measure the oscillation of muon neutrinos produced at the J-PARC accelerator in Japan. The Super-Kamiokande detector acts as the far detector and a near detector is installed in the path of the neutrino beam to reduce uncertainties on the flux and cross sections. This near detector, called ND280, is located 280 m from the production target of the T2K neutrino beamline in order to measure the characteristics of the beam prior to any oscillation. The first ND280 analysis of the momentum-angle distribution of the muon produced by muon neutrino charged current interactions has been performed. These results have also been used to determine a flux-averaged cross section.

1. Introduction

The T2K experiment was designed to measure the oscillation of muon neutrinos produced at the J-PARC facility in Tokai, Japan. The disappearance of muon neutrinos and appearance of electron neutrinos are measured in the Super-Kamiokande detector (SK) acting as the far detector. The T2K complex also includes near detectors located 280 m from the production target. One of these detectors is ND280, which is located in the path of the neutrinos going to Super-Kamiokande. The role of ND280 is to measure the characteristics of the unoscillated neutrino beam such as the energy spectrum and also the interaction rate.

The main purpose of the ND280 muon neutrino measurement is to constrain the flux and cross section parameters of the T2K neutrino oscillation analysis in order to reduce systematic uncertainties. The ND280 results can also be used to measure the differential cross section to test the cross section models used by T2K and other neutrino experiments.

2. The apparatus

ND280 is a complex of multiple subdetectors all installed inside the magnet from the UA1 experiment, which provides a 0.2 T magnetic field [1] (Fig. 1). The subdetectors used in the analysis presented in this paper are the two fine-grained detectors (FGDs) [3] and the three time projection chambers (TPCs) [2] which together are referred to as the tracker. The FGDs are composed of alternating vertical and horizontal layers of 1 cm^2 square extruded polystyrene scintillator bars read out by wavelength-shifting fibers. The purpose of the FGDs is to act as active targets and provide detailed vertex information of the neutrino interactions. FGD1 is composed entirely of scintillator layers while FGD2 contains scintillator and water layers in order to compare the neutrino interaction rate on carbon and on oxygen. FGD1 and FGD2 are

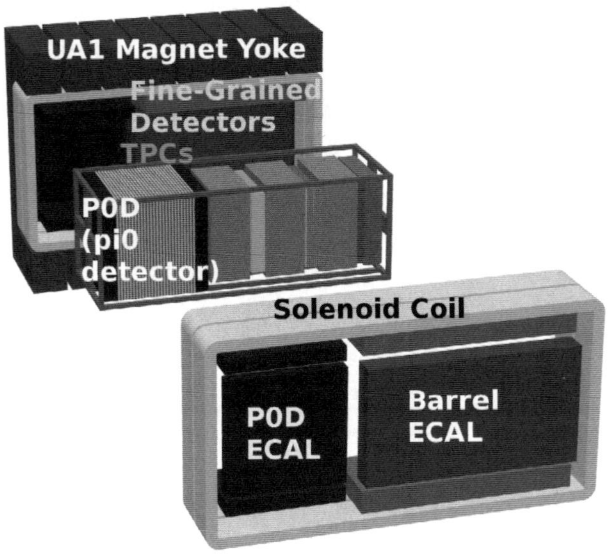

Figure 1. Schematics of ND280 in the opened magnet position.

located respectively between TPC1 and TPC2, and TPC2 and TPC3. The TPCs are filled with an argon, CF_4, and isobutane gas mixture at respectively 95, 3, and 2% and use MicroMegas detectors for gas amplification with pad readout. The TPCs are used to reconstruct the charged particle's momentum with an inverse momentum resolution of 0.1 $(\text{GeV}/c)^{-1}$. The TPCs are also capable of particle identification using the energy loss to distinguish in particular muons from electrons with a misidentification probability of $< 0.9\%$ from 200 MeV/c to 1.8 GeV/c. ND280 also contains, upstream of the tracker, the P0D which is dedicated to π^0 reconstruction and made of scintillator bars interleaved with lead and brass sheets. The tracker and the P0D are surrounded by electromagnetic calorimeters (ECals) composed of layers of plastic scintillator bars with lead sheets in between.

3. The ND280 muon neutrino measurement

The analysis of the ND280 data to extract the muon neutrino beam properties starts by selecting all the charged current (CC) interactions occurring in the FGD1 detector. In each event, the highest-momentum negative track is the muon candidate track. The selected events have the following requirements: they have no track in TPC1, their muon candidate track originates from the FGD1 fiducial volume, and the particle identification from the TPCs for the candidate track is consistent with the muon hypothesis. This CC interaction sample is split into a charged current quasi-elastic (CCQE) sample, and a charged current non quasi-elastic (CCnQE) sample for interactions in which at least one pion was produced. Events in the CCQE sample have only one TPC-FGD track and no Michel electron in FGD1, originating from the decay of a pion stopped in FGD1 producing a muon, and then an electron. The Michel electrons are identified using timing information in FGD1.

The results presented in this paper use only the data from 2010 and 2011, however the consistency of these data with the 2012 data was tested and confirmed. Momentum-angle spectra are produced for the CCQE and CCnQE samples. The main physics application of these results is to estimate beam and neutrino interaction properties with a maximum likelihood fit of the ND280 data along with the beam flux prediction and external cross section data as constraints. The results of the fit and the high correlation between the ND280 and SK fluxes provide a significant reduction of the beam and cross section systematic uncertainties in the SK neutrino interaction rate prediction. The CCQE and CCnQE samples were split in the ND280 analysis to allow the CCQE sample to constrain the spectral shape, the flux and cross section parameters, while the CCnQE sample constrains backgrounds and cross section parameters. The momentum projection of the CCQE and CCnQE spectra prior to the fit (Fig. 2 and 3) display already a good agreement between data and MC.

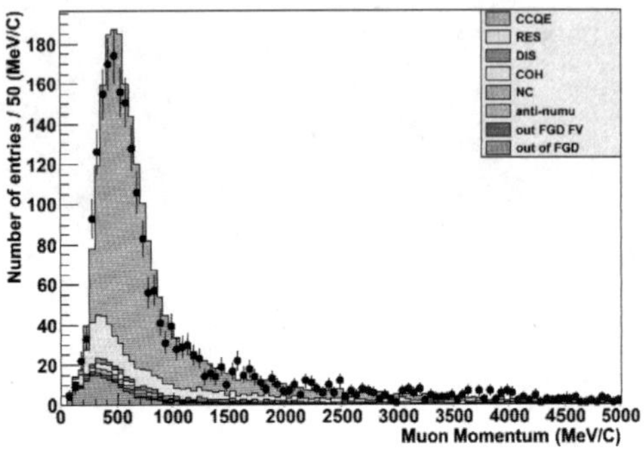

Figure 2. Momentum distribution of muon from selected CCQE interactions for data and simulation.

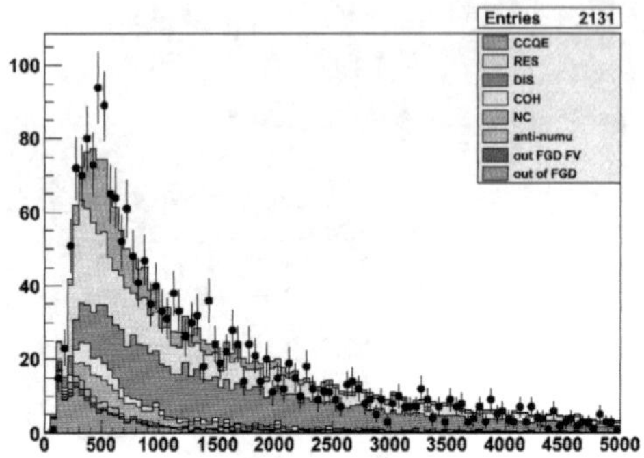

Figure 3. Momentum distribution of muons from selected CCnQE interactions for data and simulation.

4. Cross section measurement

In addition to constraining flux and cross section parameters for the neutrino oscillation measurement, the ND280 results can be used to perform a flux-averaged differential cross section measurement.

For this analysis, the CC inclusive sample containing both the CCQE and CCnQE samples is used to have a data sample as large as possible. The cross section is measured from the momentum-angle spectrum using:

$$\langle\frac{\partial^2\sigma}{\partial p_\mu \partial\cos\theta_\mu}\rangle_{kl} = \frac{N_{kl}^{\text{int}}}{T\phi\Delta p_{\mu,k}\Delta\cos\theta_{\mu,l}} \tag{1}$$

with N_{kl}^{int} the number of true interactions in the true bin kl, T the number of target nucleons, ϕ the flux, p_μ the muon momentum, and $\cos\theta_\mu$ the angle between the muon direction and the neutrino beam axis. In order to obtain N_{kl}^{int}, we unfold the momentum-angle resolution from the measured CC inclusive spectra using an iterative method based on the Bayes' theorem [4]. A migration matrix derived from the simulation converts the reconstructed bins into true bins and a correlation matrix is used to propagate the systematic uncertainties of each bin. The measured differential cross section is compared in Fig. 4 to the predictions from the two neutrino interaction generators used in T2K called NEUT [5] and GENIE [6]. The integrated cross section is:

$$\langle\sigma_{\text{CC}}\rangle_\phi = (6.93 \pm 0.13(stat.) \pm 0.85(syst.)) \times 10^{-39}\text{cm}^2/\text{nucleons}. \tag{2}$$

Similarly the predictions from NEUT and GENIE are respectively:

$$\langle\sigma_{\text{CC}}^{\text{NEUT}}\rangle_\phi = 7.26 \times 10^{-39}\text{cm}^2/\text{nucleons}, \tag{3}$$

$$\langle\sigma_{\text{CC}}^{\text{GENIE}}\rangle_\phi = 6.68 \times 10^{-39}\text{cm}^2/\text{nucleons}. \tag{4}$$

The measurement is limited by the flux and detector systematic uncertainties. Future flux determination will use the 2009 data set from the NA61 experiment, which used a graphite target similar to the one used in the T2K beamline [7], and therefore the flux uncertainties will be reduced. Also improvements in the ND280 software will provide better track reconstruction and reduced detector systematic uncertainties.

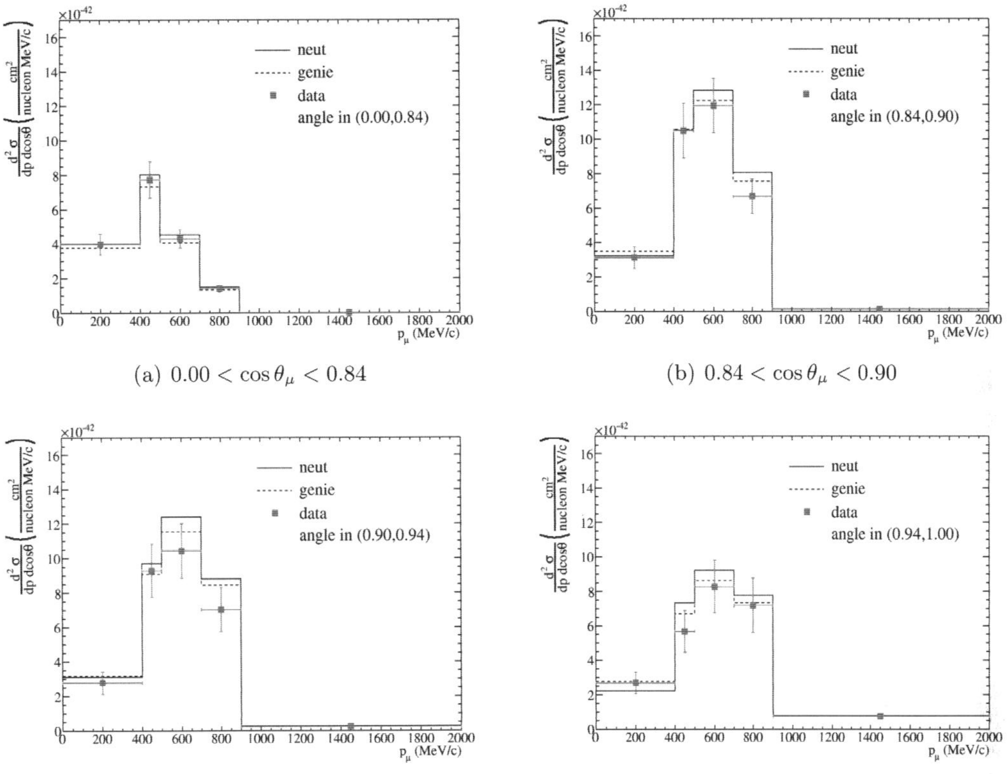

(a) $0.00 < \cos\theta_\mu < 0.84$ (b) $0.84 < \cos\theta_\mu < 0.90$

(c) $0.90 < \cos\theta_\mu < 0.94$ (d) $0.94 < \cos\theta_\mu < 1.00$

Figure 4. Flux-averaged differential cross section in momentum of the outgoing muon for each angle bin, for data and for the two neutrino interaction generators NEUT and GENIE.

5. Future analysis developments

New extensions of the ND280 analysis are already being developed in order to increase the constraints on the flux and cross section parameters used in the oscillation measurement. The first approach is to use the muon and the proton tracks in the CCQE interactions instead of just identifying the muon track. A more detailed study of the CCQE interaction will be possible starting with the neutrino energy, which will be reconstructed from both particles, eliminating the need for assumptions on the neutron kinematics. This analysis will also have a stronger discrimination between neutrino interaction models since it will test the modelling of both the muon and the proton.

The other improvement will be the measurement of the CC1π interaction, which is a significant background of the CCQE measurement in SK due to the pion absorption. Furthermore this measurement will complement the MiniBooNE data measured for neutrinos around 1 GeV of energy. This analysis starts from the CCnQE sample and uses additional selections to identify events with one pion. The TPC particle identification is used when the pion reaches the TPCs. The possibility to identify the pions in the FGDs or the ECals is being investigated for event topologies where the pion did not reach a TPC.

6. Conclusion

The T2K near detector has demonstrated its important role in the neutrino oscillation measurement by providing momentum-angle distributions for CCQE and CCnQE interactions. The results from ND280 have also been used for a stand-alone cross section measurement of CC interactions. The current results correspond only to a few percent of the statistics goal of the T2K experiment. Furthermore calibration, software, and analysis improvements have already been implemented and more are in development.

References

[1] Abe K *et al.* 2011 *Nucl. Instrum. Methods* A **659** 106
[2] Abgrall N *et al.* 2011 *Nucl. Instrum. Methods* A **637** 25
[3] Amaudruz P-A *et al.* 2012 *Nucl. Instrum. Methods* A **696** 1
[4] D'Agostini G 1995 *Nucl. Instrum. Methods* A **362** 487
[5] Hayato Y 2009 *Acta Phys. Pol.* B **40** 2477
[6] Andreopoulos C *et al.* 2010 *Nucl. Instrum. Methods* A **614** 87
[7] Abgrall N *et al.* 2012 *Nucl. Instrum. Methods* A **701** 99

ICARUS-NESSiE: a sensitive search for sterile neutrinos at CERN SPS

A. Guglielmi [1]

INFN, Sezione di Padova and Department of Physics and Astronomy "Galileo Galilei", via Marzolo 8 35131, Padova, Italy

E-mail: alberto.guglielmi@pd.infn.it

Abstract.
A new experimental search for sterile neutrinos beyond the Standard Model at a new CERN-SPS neutrino beam aiming at measuring the electron and muon neutrino events with a Near and Far detectors (1600 and 330 m from the proton target) is presented. The project will exploit the ICARUS T600 LAr-TPC moved from LNGS to the CERN Far position and a new additional LAr-TPC detector, 1/4 of the T600, located in the Near position. Two magnetic spectrometers will be placed downstream of the two LAr-TPC detectors to greatly complement the physics capabilities. Comparing the two detectors, in absence of oscillations, all cross sections and experimental biases cancel out. Any difference of the event distributions at the two locations should be attributed to the possible existence of oscillations, presumably due to additional neutrinos with a mixing angle $\sin^2(2\theta_{new})$ and a mass squared difference Δm^2_{new} larger than the measured for the standard neutrinos. The superior quality of the LAr imaging TPC, in particular its unique electron-π^0 discrimination allows for full rejection of backgrounds and offers a lossless ν_e detection capability. The determination of the muon charge with the spectrometers allows for the full separation of ν_μ from $\overline{\nu}_\mu$ and therefore controlling systematics from muon mis-identification mainly at high momenta.

1. Introduction

The discovery of a Higgs boson at LHC CERN collider has crowned the Standard Model (SM) of fundamental interactions calling also for a verification of the Higgs coupling to the gauge boson and to fermions.

Neutrinos, being the only elementary fermions whose basic properties are still largely unknown, must naturally be one of the main priorities to complete our knowledge of the SM. Moreover the discovery of neutrino oscillations and of the related non zero masses represents today a main experimental evidence of new physics. Albeit still unknown precisely, the incredible smallness of the neutrino rest masses, compared to those of other elementary fermions points to some specific scenario, awaiting to be elucidated. On the other hand, the astrophysical importance of neutrinos is immense: if heavy enough, they may also contribute to dark matter in the Universe.

The possible presence of sterile neutrinos which do not fully see the ordinary electro-weak interactions but still oscillating and mixing with ordinary neutrinos, was originally proposed by B. Pontecorvo in a seminal paper where he considered the existence of right-handed neutrinos,

[1] On behalf of the ICARUS NESSiE Collaborations

Journal of Physics: Conference Series **460** (2013) 012014 doi:10.1088/1742-6596/460/1/012014

neutrino oscillations, lepton number violation, $0\nu - \beta\beta$ decays, and other fundamental queries that have dominated neutrino physics up to now [1].

Two distinct classes of anomalies have been so far reported, although not at an entirely conclusive level, The apparent 3σ disappearance of the $\overline{\nu}_e$ low energy neutrinos events from nuclear reactor experiments [2] and from the signal from Mega-Curie calibration sources in GALLEX and Sage experiments ($R = 0.86 \pm 0.05 \sim 2.7\sigma$ away from R = 1 expectation) [3] could originate from ν_e ($\overline{\nu}_e$) conversion into "invisible" sterile components, leading to the observation of oscillatory, distance dependent, disappearance rates.

The second anomaly refers to the evidence of an electron excess originated by initial ν_μ beam from particle accelerators as observed in LSND (3.8 σ) [4] and MiniBooNe (3 σ) [5] experiments where some distance dependent $\nu_\mu \rightarrow \nu_e$ oscillations may be observed as a ν_e excess, especially in the antineutrino channel.

Moreover the possible existence of additional neutrino states may be also hinted – or at least not excluded – by cosmological data [6].

All the recalled "anomalies", having accumulated an impressive number of standard deviations, may indeed hint at an unified scheme in which the values of Δm^2_{new} may have a common origin, the different values of $\sin^2 2\theta_{new}$ for different channels reflecting the structure of $U_{(4,k)}$ matrix (or even with an higher number of neutrinos) with $k = \mu$ and e. This long standing series of experimental results hinting at the hypothesis of sterile neutrinos deserve a major effort to definitely establish the existence of additional neutrino states and clarify the underlying physics.

The proposed experiment at a new CERN SPS neutrino beam may be in a unique position to be able to unambiguously investigate with high sensitivity such an oscillatory scenario by the simultaneous measurement of the neutrino fluence at two different positions. For these purposes two large mass LAr-TPC's, ICARUS T600 presently taking data at LNGS Laboratory and a new additional smaller T150 LAr-TPC detector will be settled with two magnetic spectrometers placed downstream of the two LAr-TPC's to greatly complement the physics capabilities. Only in this way the new values of Δm^2 and of $\sin^2 2\theta$ can be separately identified. All other experimental non accelerator programs presently under investigation will not focus with an equal sensitivity on the direct parameters of the oscillation phenomenon.

2. ICARUS-T600: a powerful detector for neutrino physics

ICARUS-T600 detector, the largest LAr-TPC ever built [7], consists of two identical adjacent cryostats, filled with 770 t of ultra-pure liquid Argon corresponding to 470 t of imaging mass, each one housing 2 TPCs separated by a central common cathode, with 1.5 m drift length. Ionization electrons, abundantly produced by charged particles along their path are drifted by an uniform electric field (E_D = 500 V/cm) towards the TPC anode made of 3 parallel wire planes (3 mm pitch), facing the drift volume, oriented at 0^0, $\pm 60^0$ with respect to the horizontal direction. The relative time of each ionization signal, combined with the electron drift velocity information ($v_D \sim 1.6$ mm/μs), provides the position of the track along the drift coordinate. Globally a remarkable 1 mm^3 spatial resolution has been measured. Absolute event time is determined exploiting the scintillation light prompt signal from arrays of photo-multipliers installed in LAr behind the wire planes.

ICARUS is a very suited detector for rare event studies, such as neutrino oscillation physics from CNGS beam and atmospheric neutrinos, and search for nucleon decay, since it combines large mass with good spatial granularity and calorimetric accuracy. The calorimetric response ($\sigma_E/E \approx 11\%/\sqrt{\mathrm{E(GeV)}}$) allows a precise measurement of energy for contained events, while the momentum of non-contained particles (mainly muons) can be measured within $\sim$ 20% resolution by multiple scattering from the deflection angle along the track. At higher energies the estimated resolution for hadronic showers is $\sigma_E/E = 0.30/\sqrt{E(GeV)}$. The measured dE/dx

energy loss allows a clear pion to proton separation as obtained from the last part of the residual range. Moreover $\pi^0 \to 2\gamma$ are recognized by detecting γ conversion, measuring dE/dx and the 2γ invariant mass (radiation length in LAr is $X_0 = 14$ cm). As a result a very efficient e/π^0 discrimination allows to reject the neutrino NC interactions by a factor $\approx 10^3$ in the ν_e appearance search, while keeping more than 90% of ν_eCC events. A typical ν_μ CC event, fully reconstructed, is shown in Fig. 1.

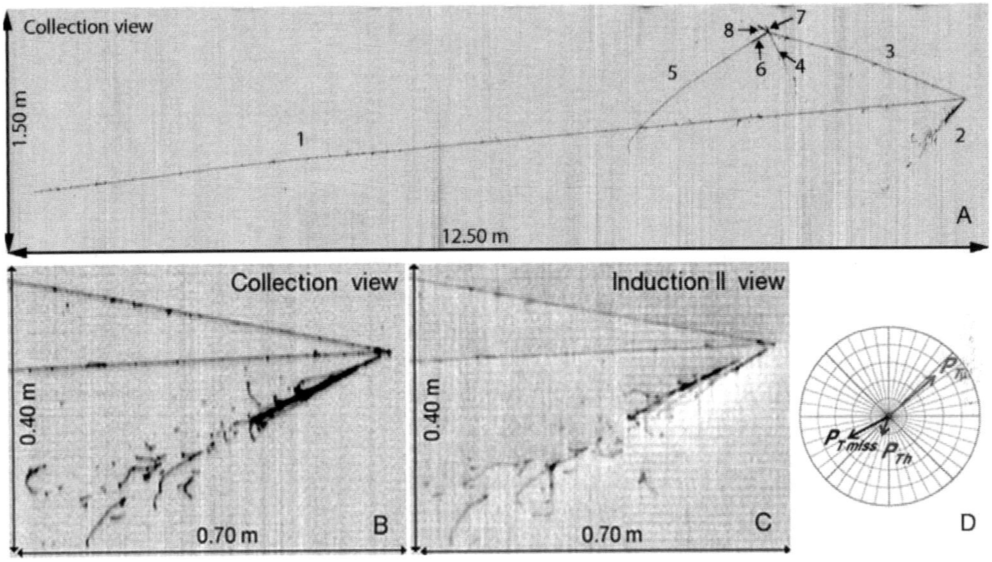

Figure 1. A reconstructed ν_μCC event in ICARUS. The neutrino produces a long muon (1), a charged pion (3) and an e.m. shower (2). At a closer look the shower can be identified as generated by a π^0; two e^-e^+ pairs can be resolved, especially in Induction2 view, and the measured invariant mass of the 2γ system $M^*_{\gamma\gamma} = 125 \pm 15$ MeV is consistent with the π^0 mass. The momentum of the escaping muon measured by multiple scattering, is 10.5 ± 1.8 GeV. In the secondary vertex, the π interaction produces several hadrons; a $K \to \mu \to e$ is clearly recognized (5). The 250 MeV/c total p_T is consistent with the nucleon Fermi momentum.

The highly efficient data taking at LNGS Laboratory with CNGS beam neutrino beam allowed to record an event statistics of $8.6 \cdot 10^{19}$ pot during 2010-2012 runs, triggering on the photomultiplier signals in coincidence with the early-warning signal sent from CERN at each proton extraction. Based on 1091 neutrino events, $\sim$ 50% of the ICARUS data collected in 2010-2011, the ICARUS Collaboration reported an early result on the search for a $\nu_\mu \to \nu_e$ signal due to the LSND anomaly in the CNGS beam at an average energy of about 20 GeV, after a flight path of $\sim$730 km [8]. The LSND anomaly would manifest as an excess of ν_e events, characterized by a fast energy oscillation averaging approximately to $\sin^2(1.27\Delta m^2_{new} L/E_\nu) \approx 1/2$ with probability $P_{\nu_\mu \to \nu_e} = 1/2 \sin^2(2\theta_{new})$. Two clear ν_e events have been found, compared with the expectation of 3.7 ± 0.6 events from conventional sources. The result, compatible with the absence of a LSND anomaly sets a 90% and 99% confidence level limits for the oscillation probabilities $\langle P_{\nu_\mu \to \nu_e} \rangle \leq 5.4 \times 10^{-3}$ and $\langle P_{\nu_\mu \to \nu_e} \rangle \leq 1.1 \times 10^{-2}$ respectively (Fig. 2). Moreover this result strongly limits the window of open options for the LSND anomaly to a narrow region around $(\Delta m^2, \sin^2(2\theta)) = (0.5 \text{ eV}^2, 0.005)$, where a 90 % overall agreement between the present ICARUS measurement, the published limits of KARMEN and the published positive signals of LSND and MiniBooNE Collaborations, is found.

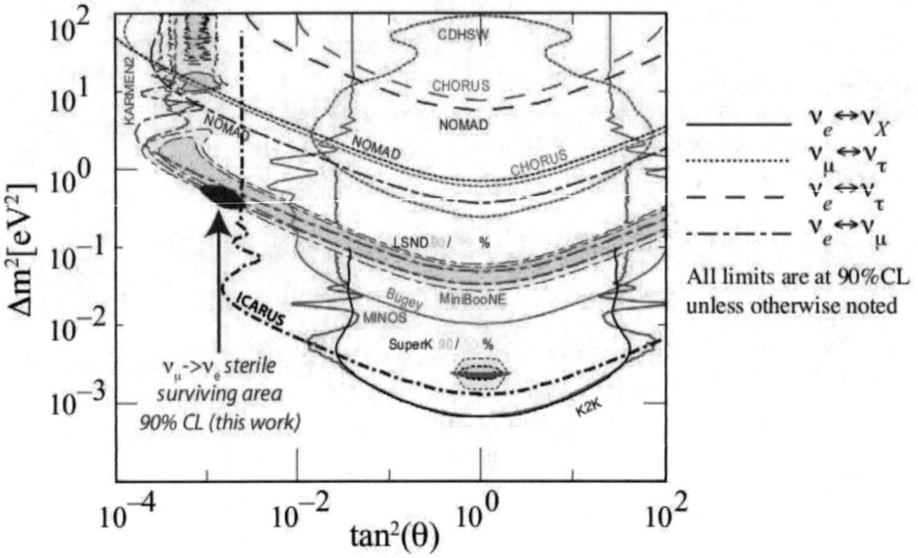

Figure 2. Regions in the $(\Delta m^2, \tan^2(\theta))$ plane excluded by the ICARUS experiment compared with the published results. While for $\Delta m^2 \gg 1eV^2$ there is a disagreement for $\nu_\mu \to \nu_e$ between the allowed regions from the published experiments, for $\Delta m^2 \leq 1eV^2$ the ICARUS result now allows to define a much smaller, narrower allowed region centered around $(\Delta m^2, \sin^2(2\theta)) = (0.5eV^2, 0.005)$ in which there is a 90% C.L. overall agreement.

3. The ICARUS-NESSiE experiment

In order to definitely clarify the experimental hints for the presence of sterile neutrinos, exploring in particular the surviving LSND/MiniBooNe $(\sin^2 2\theta,\ \Delta m^2)_{new}$ parameter region as indicated by ICARUS measurement, the proposed experiment will operate at CERN SPS at much shorter distances with neutrino energies lower than in the CNGS beam line.

The experimental set-up [9] is based on two identical LAr-TPC's [10] complemented by magnetized spectrometers [11] detecting electron and muon neutrino events at Far and Near positions, 1600 m and 330 m respectively from the proton target in the CERN North Area (Figure 3). A new intense 100 GeV/c fast extracted proton beam, 4.5×10^{19} pot/year, will produce a high intensity on-axis ν beam with an energy spectrum centered around $E_\nu \sim 2$ GeV (Figure 4).

More precisely the proposed experiment will provide:

- L/E oscillation path-lengths appropriate to match to the Δm^2 window for the expected anomalies;
- "imaging" detectors capable to identify unambiguously all reaction channels with a *"Gargamelle class"* LAr-TPC;
- magnetic spectrometers able to determine muon charge with few % mis-identification and momentum in a broad range;
- a clean measure of interchangeable ν and anti-ν focused beams;
- very high event rates ($> 10^6\nu_\mu$ and $\simeq 10^4\nu_e$) due to large detector masses, allowing to record relevant effects at the % level;
- initial ν_e and ν_μ components cleanly identified.

Journal of Physics: Conference Series **460** (2013) 012014 doi:10.1088/1742-6596/460/1/012014

The project will exploit the ICARUS T600 detector, now running in the LNGS underground laboratory, moved at the CERN Far position. An additional 1/4 of the T600 detector (T150), a LAr-TPC identical to one module of T600 but a factor 2 reduced length, will be constructed and located in the Near position. Two spectrometers will be placed downstream of the two LAr-TPC detectors to greatly complement the physics capabilities. Spectrometers will exploit a classical dipole magnetic field with iron slabs, and a new concept air-magnet, to perform charge identification and muon momentum measurements in a wide energy range from sub-GeV to several GeV, over a transverse area larger than 50 m^2 with about 2500 m^2 of RPC precision trackers.

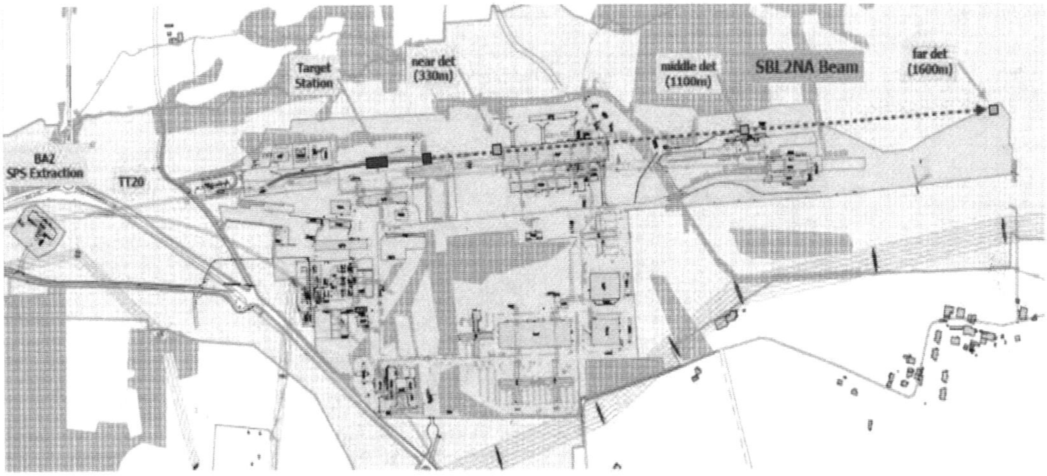

Figure 3. The new SPS North Area neutrino beam layout. Main parameters are: primary beam: 100 GeV; fast extraction from SPS; target station next to TCC2, ~11 m underground; decay pipe: 100 m, 3 m diameter; beam dump : 15 m of Fe with graphite core, followed by muon stations; neutrino beam angle: pointing upwards at ~3 m in the far detector ~5 mrad slope.

At the two positions the energy spectra of the ν_e beam component are expected to be coincident but for second order effects which can be reliably reproduced (Figure 4). In absence of oscillations, since all cross sections and experimental biases cancel out, the observed event distributions at the near and far detectors must be the same. Any difference can be ascribed to the possible existence of ν-oscillations due to additional neutrinos with new mixing angles $\sin^2 2\theta_{ij}$ and mass differences Δm^2_{ij} larger than those measured in the standard three neutrino family scheme.

The superior quality of the LAr imaging TPC, now widely experimentally demonstrated, and in particular its unique *electron* - π^0 discrimination, allows full rejection of backgrounds with a lossless ν_e detection capability. The determination of the muon charge with the spectrometers allows the full separation of ν_μ from $\overline{\nu}_\mu$ and an improved control of systematics from muon mis-identification.

Compared to the ICARUS measurement at the CNGS beam, the much shorter distances and lower neutrino energies will allow to increase the event rate, reduce the overall particle multiplicity of the neutrino events, and enlarge the angular range of the secondaries substantially improving the ν_e selection efficiency. A 90% electron detection efficiency including the event energy reconstruction is expected inside a fiducial volume defined by $4X_0$ and $0.7X_0$ longitudinal and side cuts respectively. This will ensure a 90% ν_e detection efficiency while keeping the mis-interpretation probability for π^0 from NC interaction at the 0.1% level inside the fiducial volume.

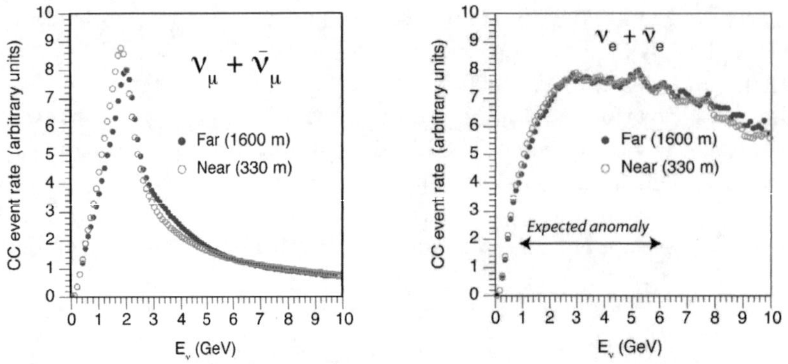

Figure 4. Muon (left) and electron (right) neutrino CC interaction spectra, at near and far positions.

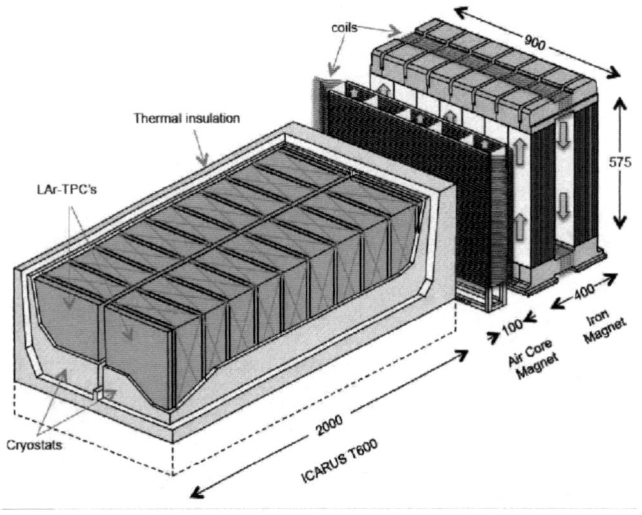

Figure 5. ICARUS LAr-TPC and NESSiE spectrometer at far site.

All the previously described anomalies will be explored with both neutrino and anti-neutrino focused beams. The separation of ν_μ from $\overline{\nu}_\mu$ will allow to exploit the interplay of the different possible oscillation scenarios, as well as the interplay between disappearance and appearance of different neutrino states and flavors. Moreover the NC/CC ratio will provide a sterile neutrino oscillation signal by itself and it will beautifully complement the normalization and the systematics studies. This experiment will offer remarkable discovery potentialities, collecting a very large unbiased event statistics (order of millions) both in the neutrino and antineutrino channels, largely adequate to definitely settle the origin of the ν-related anomalies.

The experiment is sensitive to $sin^2 2\theta$ down to 3×10^{-4} (for $|\Delta m^2| > 1.5eV^2$) and $|\Delta m^2|$ down to $0.01eV^2$ (for $sin^2 2\theta = 1$) at 90% C.L. for the $\nu_\mu \rightarrow \nu_e$ transition with one year exposure (4.5×10^{19} pot). The parameter space region allowed by the LSND experiment is fully covered, except for the highest Δm^2 region (Figure 6 left). The sensitivity has been computed according to the above described particle identification efficiency and assuming a 3% systematic uncertainty in predicting the Far to Near ν_e ratio. A further control of the overall systematics

Journal of Physics: Conference Series **460** (2013) 012014 doi:10.1088/1742-6596/460/1/012014

will be provided by the LAr and spectrometer combined measurement of CC spectra in the near site and over the full energy range. In anti-neutrino focusing, twice as much exposure (0.9×10^{20} pot) allows to cover both the LSND region and the new MiniBooNE results (Figure 6 right). Both favored MiniBooNE parameter sets, corresponding to two different energy regions in the MiniBooNE antineutrino analysis, fall well within the reach of this proposal. In Figure 7 left the sensitivity for ν_e disappearance search in the $sin^2(2\theta)$, Δm^2 plane is shown for one year data taking. The oscillation parameter region related to the anomalies from the combination of the published reactor neutrino experiments, GALLEX and SAGE calibration sources experiments is completely explored. The ν_μ disappearance signal is well studied by the spectrometers, with a very large statistics, and disentangling of ν_μ and $\overline{\nu}_\mu$ interplay. Figure 7 right shows the sensitivity plot (at 90% C.L.) for two years negative-focusing plus one year positive-focusing. SPSC-P347 A large extension of the present limits for ν_μ by CDHS and the recent SciBooNE+MiniBooNE will be achievable in $sin^2(2\theta)$, Δm^2.

The ICARUS-T600 shutdown at LNGS will start in the middle 2013 with the LAr emptying phase, followed by the dismantling of the detector which components will be transported to CERN to be reassembled in a dedicated hall before being moved to the Far site in the North Area. The T150 detector of the Near site will be realized closely following the T600 specifications. Some refurbishing of the photomultiplier system, DAQ electronics and trigger system will be realized with the aim of increasing the detector performance. The different components of the NESSiE spectrometers are expected to be separately realized and then installed in the Far and Near pits downstream the LAr-TPCs. According to the CERN plan the new short baseline neutrino facility at SPS North Area is expected to be operational early 2016 with the neutrino beam commissioning followed by the data taking.

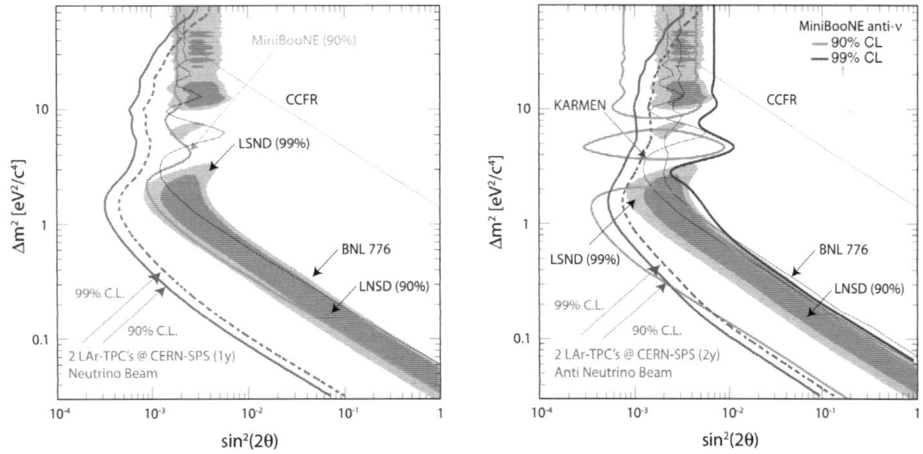

Figure 6. Expected muon to electron neutrino oscillation sensitivity for the proposed experiment exposed at the CERN-SPS neutrino beam (left) and anti-neutrino (right) for 4.5×10^{19} pot (1 year) and 9.0×10^{19} pot (2 years), respectively.

[1] Pontecorvo B 1967 *Zh. Eksp. Teor. Fiz.* **53** 1717 [1968 *Sov. Phys. JETP* **26** 984]CosmoArg
[2] Mention G et al. 2011 *Phys.Rev.* D **83** 073006 and previous references therein
T. Lasserre, Talk to this Symposium, Paris (France)
[3] Abdurashitov J N et al. (SAGE Coll.) 2009 *Phys. Rev.* C **80**, 015807
Abdurashitov J N et al. (SAGE Coll.) 1996 *Phys. Rev. Lett.* **77** 4708
Abdurashitov J N et al. (SAGE Coll.) 1999 *Phys. Rev.* C **59** 2246
Abdurashitov J N et al. (Sage Coll.) 2006 *Phys. Rev.* C **73** 045805

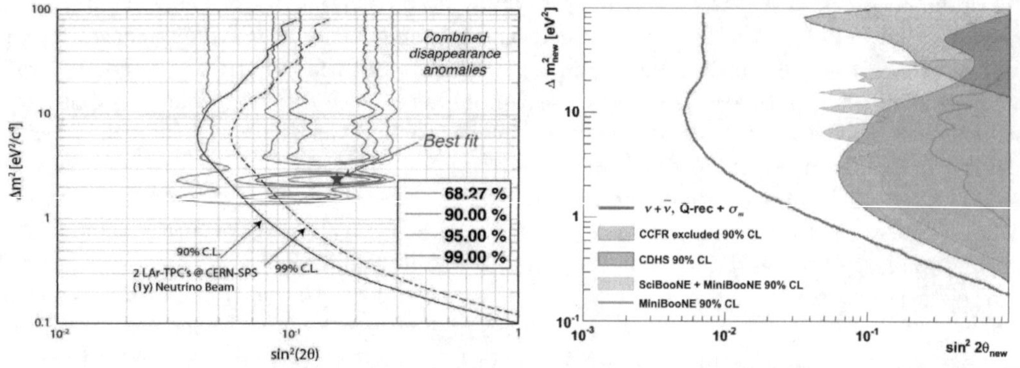

Figure 7. Sensitivity for ν_e disappearance for 1 year data taking (left) and for ν_μ disappearance considering 3 years of the CERN-SPS beam (2 years in anti-neutrino and 1 year in neutrino mode) from CC events fully reconstructed in NESSiE+LAr.

Kaether F, Hampel W, Heusser G, Kiko J, and Kirsten T 2010 *Phys. Lett.* B **685** 47
Anselmann P et al. (GALLEX Coll.) 1995 *Phys. Lett.* B **342** 440
Hampel W et al. (GALLEX Coll.) 1998 *Phys. Lett.* B**420** 114
[4] A. Aguilar A. et al. (LSND Coll.) 2001 *Phys. Rev.* **D 64** 112007
[5] Aguilar-Arevalo A A (MiniBooNE Coll.) 2009 *Phys. Rev. Lett.* **102** 101802; F.Mills, ICHEP 2010, Paris, France; R. Van de Water, Neutrino 2010, Athens, Greece; E.D. Zimmerman, PANIC 2011, Cambridge, U.S.A.; A. A. Aguilar-Arevalo (MiniBooNE Coll.), arXiv:1207.4809;
[6] Komatsu E et al. 2012 *Astrophys. Jour. Suppl.* **192** 18
Dunkley J et al. 2012 *Astrophys. Jour.* **749** 90
Hamann J et al. 2010 *Phys. Rev. Lett.* **105** 181301
Izotov Y I, Thuan T X 2010 *Astrophys. Jour. Lett.* 710 67
[7] Amerio S et al (ICARUS Coll.) 2004 *Nucl. Instr. and Meth.* A **527** 329
Rubbia C et al (ICARUS Coll.) 2011 *JINT 6* 0711
[8] Antonello M et al. (ICARUS Coll.) 2012 "Experimental search for the LSND anomaly with the ICARUS LAr-TPC detector in the CNGS beam" arXiv:1209.0122.
[9] Antonello M et al. 2012 "Search for anomalies from neutrino and anti-neutrino oscillations at $\Delta m^2 \sim 1eV^2$ with muon spectrometers and large LArTPC imaging detectors" SPSC-P-347
[10] Rubbia C et al. 2011 "A comprehensive search for anomalies from neutrino and anti-neutrino oscillations at large mass differences ($\Delta m^2 \sim 1eV^2$) with two LArTPC imaging detectors at different distances from the CERN-PS", SPSC-P-345
Rubbia C et al. 2011 "Physics Programme for ICARUS after 2012" SPSC-M773
[11] Bernardini P et al. 2011 "Prospect for Charge Current Neutrino Interactions Measurements at the CERN-PS" SPSC-P-343

Journal of Physics: Conference Series **460** (2013) 012015 doi:10.1088/1742-6596/460/1/012015

Axion-like particles: possible hints and constraints from the high-energy Universe

Pierre Brun

CEA, Irfu, Centre de Saclay, F-91191 Gif-sur-Yvette — France

E-mail: pierre.brun@cea.fr

Abstract. The high-energy Universe is potentially a great laboratory for searching new light bosons such as axion-like particles (ALPs). Cosmic sources are indeed the scene of violent phenomena that involve strong magnetic field and/or very long baselines, where the effects of the mixing of photons with ALPs could lead to observable effects. Two examples are archetypal of this fact, that are the Universe opacity to gamma-rays and the imprints of astrophysical magnetic turbulence in the energy spectra of high-energy sources. In the first case, hints for the existence of ALPs can be proposed whereas the second one is used to put constraints on the ALP mass and coupling to photons.

1. Motivations for axion-like particle searches

The Standard Model of particle physics reproduces data incredibly well [1]. Some of its foundations are however not completely understood, like for example the absence of CP violation in quantum chromodynamics (QCD). The most general QCD lagrangian includes a complex phase term which –if not exactly zero– induces CP violation. The non-observation of even a very small electric dipole moment for the neutron [2] implies that this phase is smaller than 10^{-11}. This fact looks unnatural an calls for a explanation. A possible one is given by making this phase a dynamical field which value is driven to zero by the action of its classical potential. This is made possible by the introduction of a new U(1) global symmetry which is spontaneously broken at some scale f (this is the so-called Peccei-Quinn symmetry [3]). A new particle that is called the axion is then predicted as an associated pseudo Nambu-Goldstone boson [4, 5]. In the original idea of Peccei and Quinn, f was of the order of the electroweak scale (EW), inducing a mass of $\sim$100 keV for the axion, which was quickly ruled out (see [6, 7] for details). Then it was proposed that f was much greater than the EW scale, leading to a very light and weakly interacting axion (dubbed "invisible axion"). Axions are experimentally searched through their coupling to photons, from the Sun [8, 9], assuming they make up the galactic dark matter [10] or with LASERs [11]. Beyond the case of the strong CP problem and axions, axion-like particles (ALPs) appear in many models of physics beyond the Standard Model such as string theory [12, 13, 14] as pseudo Nambu-Goldstone bosons associated to the breaking of U(1) symmetries. The properties of the associated particles are similar to that of axions, but in general their mass and coupling to photons are not related, making the corresponding parameter space larger. For the general ALP case, the interaction term with photons is

$$\mathcal{L} = -\frac{1}{4} g_{\gamma a} F_{\mu\nu} \tilde{F}^{\mu\nu} a = g_{\gamma a} \vec{E} \cdot \vec{B} \, a \ , \tag{1}$$

where $g_{\gamma a}$ is the dimensionful coupling between photons and ALPs, F is the electromagnetic tensor and a is the ALP field. On the right hand-side of Eq. 1, the $F\tilde{F}$ term is expressed as a scalar product of the photon electric field and the magnetic field, revealing the fact that ALPs can couple to photons in the presence of an external magnetic field.

In the present article, the use of natural (astrophysical) environments to search for ALPs is emphasized, like during the propagation of very high energy gamma-rays over cosmological distances, and the effect of astrophysical magnetic turbulence on high-energy photon source spectra. First the conventional view of the problem of the opacity of the Universe to gamma-rays is presented, with the discussion of a possible indication for an anomalously transparent Universe. Although the possible tensions can be solved in a conventional way, they can also be released by invoking ALPs mixing with photons. Then it is shown that this observable could be used as a signature when one tries to make a discovery, but that some uncertainties prevent from using it to derive robust constraints. It is then shown that constraints can be obtained by considering the effect of magnetic turbulence around the sources and finally some examples of constraints are given.

2. The transparency of the Universe and ALPs

2.1. The conventional view of the Universe opacity to gamma-rays

Very high energy photons (with ~TeV energies) traveling through the intergalactic medium encounter different populations of background radiations. The most numerous type of background photons belong to the Cosmic Microwave Background (CMB) and a second population is the extragalactic background light (EBL). The latter has a double bump structure, that comes from direct starlight and emission re-processed by interstellar dust in the infrared band as sketched on Fig. 1. Direct measurement of the EBL is very difficult because of foregrounds and infrared radiation by the instruments.

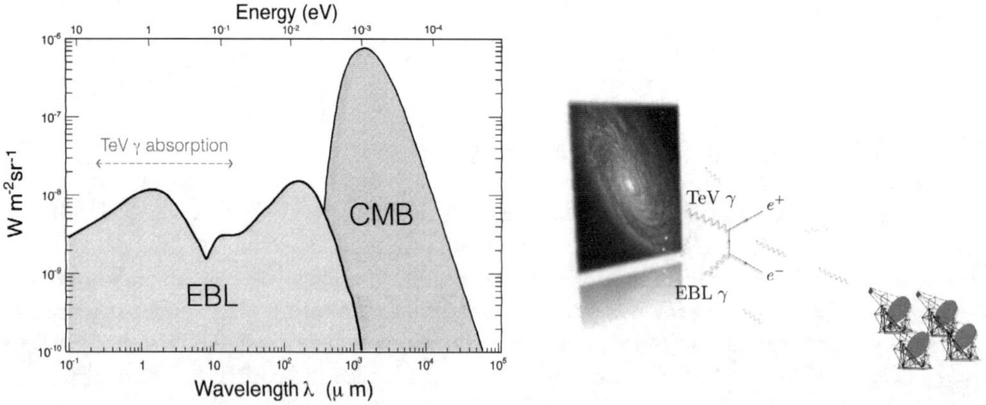

Figure 1. Left: Spectral energy density of the cosmic background photons including the CMB and the EBL (inspired from [15], with permission of the authors). Right: Illustration of the pair production process.

Journal of Physics: Conference Series **460** (2013) 012015 doi:10.1088/1742-6596/460/1/012015

TeV gamma-ray astronomy is sensitive to the EBL density and spectrum as it is responsible for the attenuation of extragalactic source fluxes at high energy. The reason for that is the pair production process $\gamma_{\rm TeV}\gamma_{\rm EBL} \to e^+e^-$, for which the threshold lies at TeV energies in the terrestrial frame. For instance considering $E_{\rm EBL} \sim 0.1$ eV, the threshold energy satisfying $E_{\rm th}E_{\rm EBL} > m_e^2$ (where m_e is the mass of the electron) yields $E_{\rm th} \sim 2.6$ TeV. In Fig. 1 the typical range of the TeV absorption range is indicated by the horizontal arrow. Because of the pair production process, the highest energy photons have a larger optical depth. Before 2006 it was commonly admitted that is was very unlikely to detect TeV photons from sources above $z \sim 0.2$. The situation changed after HESS observations of two active galactic nuclei (AGNs) at $z = 0.186$ and $z = 0.165$. As reported in [16], when unfolded from the EBL effect, the intrinsic spectra of these sources were found too hard and in tension with the source models. This was the first indication for a Universe slightly more transparent than expected at high energy. Later, AGNs were observed at redshifts as high as 0.536 by MAGIC [17] and possibly 0.61 by HESS [18]. The Universe is indeed more transparent to gamma-rays that expected. That puzzle has conventional solutions, it could be for instance that spectra are actually harder, this can be realized for instance including hadronic components in the AGN jets or in relativistic shock acceleration models. The tension can be removed as well with a revision of the EBL models, mainly with a lower density as for instance in the model of [19] which is compatible with all observations. TeV observations are now even used to provide not only upper limits on the EBL density but actual measurements [20]. The current situation is illustrated in Fig. 2, extracted from [21, 22]. It represents the energy above which the absorption becomes significant (defined by a optical depth $\tau(E) = 1$) as a function of the redshift of the source. The lines correspond to models or lower limits for the EBL density and redshift evolution from [19, 23, 24]. Constraints from the spectral indices of different sources are shown as arrows, and the HESS measurement corresponds to the blue band. One source seems to be in tension with the measurement. However the methods that lead to the constraint and the measurements are different as the constraints rely on spectral slope measurements and the measurement comes from the observation of features in the spectra that can be related to the EBL spectral density. A more unified approach might be necessary to get a definite answer on how strong the tension is. Note also that the Fermi collaboration did the same measurement at higher redshifts and found a good compatibility with EBL models [25].

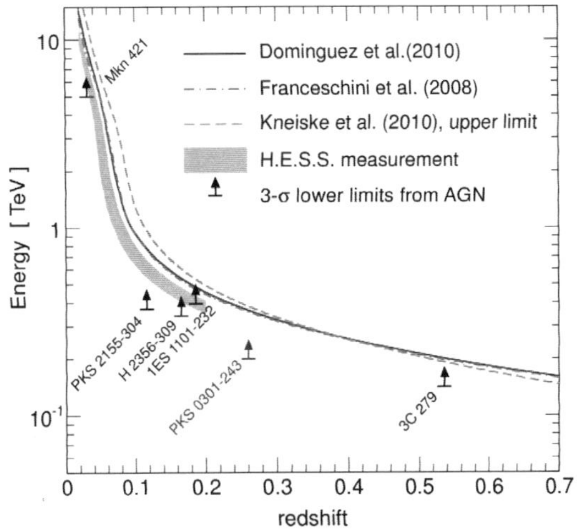

Figure 2. Energies corresponding to $\tau = 1$ for different EBL models, constraints from very high energy gamma-ray astronomy and 1-σ measurements from HESS (figure from [21, 22]).

Some studies however still claim for an anomaly, even with the lower EBL limits from [24].

It is the case in [26], where the authors claim for an anomaly, with the caveat that their claim requires to leave out some error bars.

2.2. *How ALPs enter the game*

The lack of opacity of the Universe to gamma-rays gave rise to the idea that ALPs could be responsible for this effect. The basic idea is that if mixing between ALPs and photons occur, the beam could travel in the form of ALPs on a significant fraction of way, not producing pairs, as sketched on Fig. 3. If ALPs are converted back to photons before observations, this could lead to a more transparent Universe.

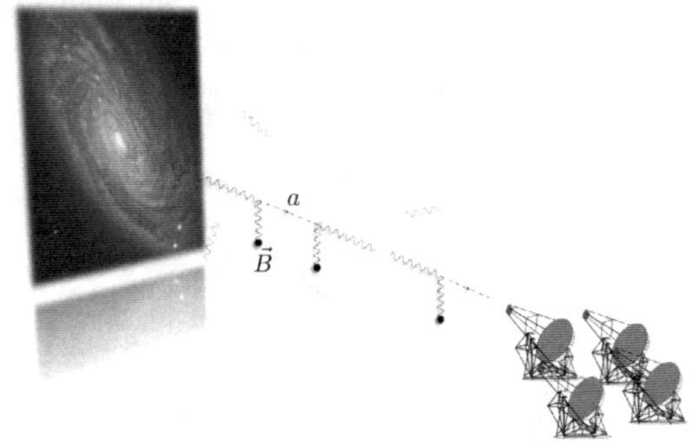

Figure 3. Illustration of the photon/ALP oscillations in a magnetic field

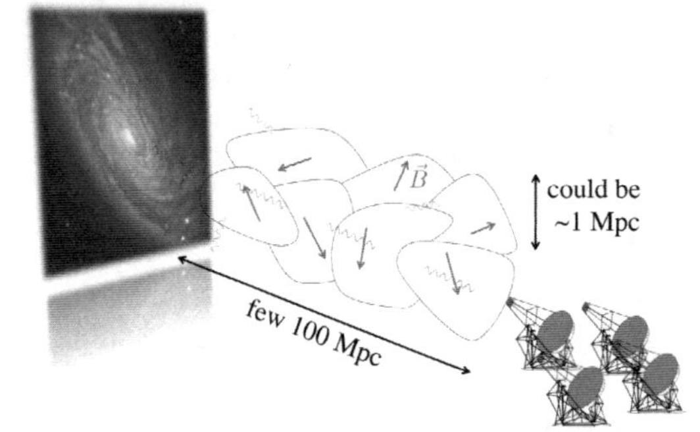

Figure 4. Illustration of the modeling of the extragalactic magnetic field

To get an idea of the relevant masses and couplings for the ALPs that are invoked here, let us consider the most simple formalism for describing the photon/ALP mixing. The system propagation is described by a Schrödinger-like equation:

$$(E - i\partial_z - \mathcal{M})\begin{pmatrix} A \\ a \end{pmatrix} = 0 \text{ with } \mathcal{M} = \begin{pmatrix} -i\frac{\tau}{2z} & \Delta_B \\ \Delta_B & \Delta_a \end{pmatrix}, \quad (2)$$

where $\Delta_B = g_{\gamma a}B_t/2$ describes the photon/ALP coupling (B_t is the transverse projection of the magnetic field), τ is the optical depth related to EBL absorption and $\Delta_a = -m_a^2/2E$ accounts for the ALP mass. The fact that the imaginary coefficient only applies to the photon part of the wavefunction leads to the change in the overall transparency. In the case of no absorption, the mixing matrix is diagonalized with a rotation angle θ such that $\tan 2\theta = -2\Delta_B/\Delta_a$. The resolution of the propagation equation in the propagation state basis leads to the probability of transition

$$P_{\gamma\to a} = \frac{1}{2}\frac{1}{1+(E_c/E)^2}\sin^2\left(\frac{g_{\gamma a}B_t\, z}{2}\sqrt{1+\left(\frac{E_c}{E}\right)^2}\right), \text{ with } E_c = \frac{m_a^2}{2g_{\gamma a}B_t}. \quad (3)$$

The overall 1/2 coefficient in $P_{\gamma\to a}$ accounts for the two polarizations of the photon. A critical energy E_c appears, that defines the energy scale at which strong mixing occurs. From the expression of E_c, with cosmological magnetic fields $B \sim 1$ nG, an ALP mass $m_a \sim$ neV and a coupling $g_{\gamma a} \sim 10^{-11}$ GeV^{-1}, the critical energy lies at the TeV scale. It follows that the type of ALPs that are concerned by the so-called transparency hint will fall in a region of low masses and with couplings larger than those of the corresponding axions.

The full treatment of the transparency problem in the presence of ALPs requires a 3×3 mixing matrix to account for the two polarization states for the photon, and a description of the magnetic field on the path from the source to the observatory. The extragalactic magnetic field is usually described as a patches of coherent domains of 1 Mpc size. The magnetic field strength is the same in all domains but from one domain to the next its orientation changes in a random way (see the sketch of Fig. 4). It can be shown (see [27]) that for random orientations and a large number N of domains, the transition probability is reduced to

$$P_{\gamma \to a} = \frac{1}{3}\left(1 - \exp\left(-3NP_0\right)\right) , \tag{4}$$

where P_0 is the transition probability in one domain. From this expression one would expect to have a 1/3 drop in the energy spectrum above E_c in the limit $NP_0 \gg 1$, and a flux that is boosted at high energy (typically above the pair-production related cutoff) as described in [28].

At least two facts lead to revise the above statements. First, in practice the limit $NP_0 \gg 1$ is hardly realized. Second, due to the unknown nature of the magnetic field configuration, the prediction on the transmission has an intrinsic variance. Indeed it can happen that the ALPs do not convert back into photons before reaching the Earth, leading in that case to an even more opaque Universe. This is nicely illustrated in Fig. 5 extracted from [29]. Here the red dot-dashed line corresponds to the conventional opacity in the absence of ALPs and the solid black line is the average prediction with ALPs. It appears that the average transparency is indeed higher than the conventional case at high energy. However, the associated uncertainty on the prediction, in other words the variance related to the randomness of the magnetic field is such that the envelope includes the conventional case. Because of that fact, if observed without ambiguity in the future, such an effect might be seen as an indication for ALP detection but could hardly serve as a firm argument for discovery.

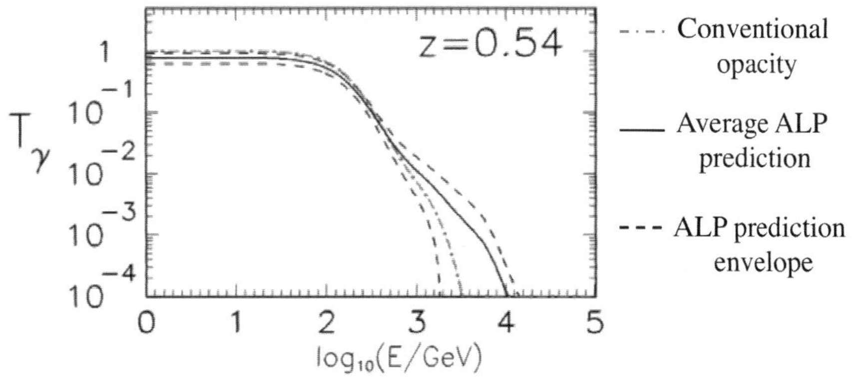

Figure 5. Transmission of photons with and without ALPs, the ALP case is ploted with the envelope corresponding to the variance on the prediction of the transparency effect (figure from [29], with permission of the authors).

Another limitation comes from the use of a very optimistic value for the extragalactic magnetic field. For the ALP effect to significantly affect the opacity, magnetic fields of nG strength with Mpc coherence length have to be present in the intergalactic medium. It is actually possible to generate such magnetic fields from inflation or QCD phase transition for instance, but the required strength is very close to current upper limits. This is illustrated in Fig. 6

(from [30]), where the different observational constraints on large scale magnetic field appear together with predictions from models (orange thin lines). There the red cross corresponds to the typical parameters used in the ALP analyses. It lies in a region that can be seen as fine-tuned given the size of the still open parameter space. At the moment it seems invoking such a strong magnetic field would be acceptable if the tension in the TeV observations was stronger.

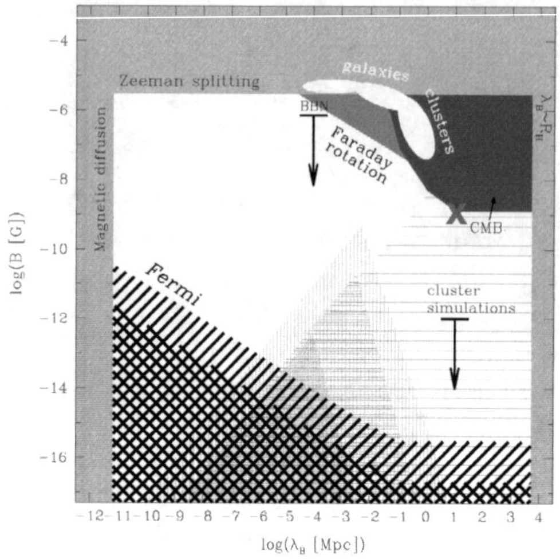

Figure 6. Constraints on large scale magnetic fields and prediction of the models. The red cross corresponds to the parameters used for the ALP solution to the transparency "hint" (figure adapted from [30], with permission of the authors).

A clever way to avoid using intergalactic magnetic fields is to remark that if the source is magnetized or embedded in a cluster, then the mixing to ALPs could occur essentially around the source. Then, the magnetic field of the Milky-Way can serve as a target magnetic field to convert back the ALPs into photons. In that case as well, a strong boost can be expected at high energy, as first proposed in [31] and then again in [32]. This effect is used in [33] to estimate possible lower limits on the $g_{\gamma a}$ coupling, but again assuming the tension is real between observations and models for the transparency. Finally it appears that the transparency observable could be used in the future for indication or discovery, if a clear tension was observed. To do so, one might wait for the next generation of gamma-ray telescopes such as CTA to have a significantly larger sample of sources.

The problem of having only a few sources can be circumvented by using an energy band for which detections are numerous. This has been proposed in [34], where the authors remark that if the strong mixing regime is realized, the statistical properties of the observed fluxes from X-ray sources could display features distinctive of ALP effects. In that case, sources would be seen with fluxes reduced by a factor of 1/3 on average. Of course having no access to the absolute intrinsic fluxes, this overall factor is not observable. However because of the random nature of the mixing process in astrophysical magnetic fields, the first and second momentum distribution should have different shapes compared to the conventional case. In [34] the authors claim the observation of anomalous features in the momenta distributions. That result has however shown to be questionable in [35] where the effect is claimed to be caused by outliers. Because in that case the detection would rely on shapes of distributions, it is difficult to infer a constraint without a deeper analysis and this effect is again used to propose a hint. Nevertheless, it illustrates one possible use of the stochastic nature of the mixing in astrophysical environments, which is no more a limitation but becomes a tool for identifying possible ALP effects. In the following, it is shown that a careful study of this randomness can lead to observable effects that are used to set constraints on the ALP parameters.

3. Constraints on ALP parameters from observations of the high-energy sky

3.1. Effect of the magnetic turbulence

One peculiar effect of photon/ALP mixing is the fact that the magnetic field turbulence can directly imprint features in the energy spectra from high-energy sources. The exact spectral shape one gets at the end is unpredictable, but as shown in [36] the statistical properties of the induced irregularities are a prediction of the ALP model. The authors of [37] and [29] already noticed that in principle the observed spectra should be very irregular in case of strong photon/ALP mixing, without considering the use of the irregularity as an observable.

To account quantitatively for the irregularity, the 2 polarizations of the photon must be considered, so that the evolution of the system after n domains is given by

$$|\psi_n\rangle = \prod_k \left(P_k^{-1}\ \exp\left[-i\left(E + \mathcal{M}_k^\star\right) s_k\right]\ P_k\right) |\psi_0\rangle\ , \quad \text{with } \mathcal{M}_k^\star = P_k \mathcal{M}_k P_k^{-1} \tag{5}$$

and

$$\mathcal{M}_k = \begin{pmatrix} -\frac{m_\gamma^2}{2E} - i\frac{\tau}{2z} & 0 & \frac{1}{2} g_{\gamma a} B_t^{(k)} \cos\phi^{(k)} \\ 0 & -\frac{m_\gamma^2}{2E} - i\frac{\tau}{2z} & \frac{1}{2} g_{\gamma a} B_t^{(k)} \sin\phi^{(k)} \\ \frac{1}{2} g_{\gamma a} B_t^{(k)} \cos\phi^{(k)} & \frac{1}{2} g_{\gamma a} B_t^{(k)} \sin\phi^{(k)} & \frac{-m_a^2}{2E} \end{pmatrix} . \tag{6}$$

k stands for the kth domain, of size s_k, P_k is the rotation matrix between the interaction eigenstates and the propagation eigenstates and the matrix $\mathcal{M}_k$ describes the mixing. The indexes k are there to recall that from one magnetic domain to the next, the corresponding parameters change due to the different orientations of the magnetic field (B_t is the projection of the magnetic field on the polarization plane and ϕ is the angle that projection makes with one of the two photon polarization). $m_\gamma = 4\pi\alpha n_e/m_e$ is the effective mass of the photon propagating in a plasma with electron density n_e. Examples of spectral oscillation patterns in one domain are given in Fig. 7 for different values of $\delta = g_{\gamma a} B_t s/2$, s being the size of the coherent domain. When several domains are considered, the spectrum ends up being very irregular as shown in Fig. 8 in the case of an unpolarized beam. For that example, an extragalactic source is considered and the magnetic field is typical of that of a galaxy cluster. The top panel of Fig. 8 is the raw signal and the bottom panel is the same signal smoothed by the energy resolution of HESS (~15%). In that case the critical energy is of order 1 TeV and the effective photon mass is negligible.

Whereas in the case of the extragalactic magnetic field the naive description of the turbulence might be sufficient (essentially because its properties are very poorly known), galaxy cluster magnetic fields may deserve a better treatment. The magnetic field in that case is modeled by a gaussian field with zero mean and a distribution of modes that is described by a Kolmogorov-like spectrum as in Eq. 7:

$$(\delta B)^2 \propto \sigma^2 \frac{k^2}{1 + (kL_c)^\gamma}\ . \tag{7}$$

The corresponding power spectrum is modeled by a function resembling that of Fig. 9. In galaxy clusters, the typical coherence length of the magnetic field is 10 kpc and the strength of the field is 1 to 10 μG.

3.2. Examples of constraints

The first example of constraints is from the HESS analysis of PKS 2155-304 [38, 39], which is an AGN located at $z = 0.116$. For that source, both the extragalactic magnetic field and the

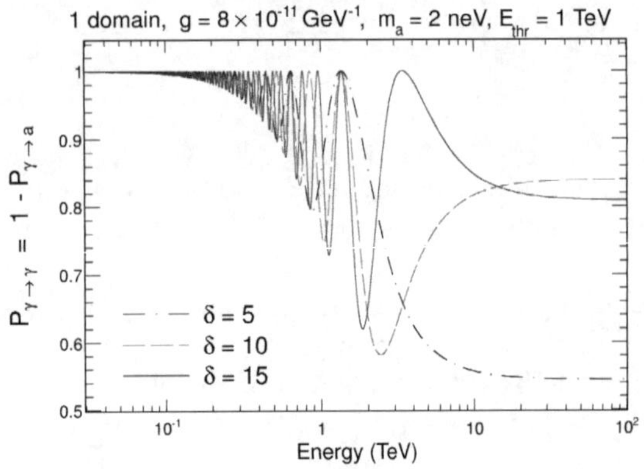

Figure 7. Spectral oscillation patterns in domains with coherent magnetic field and different ALP parameters (figure from [36]).

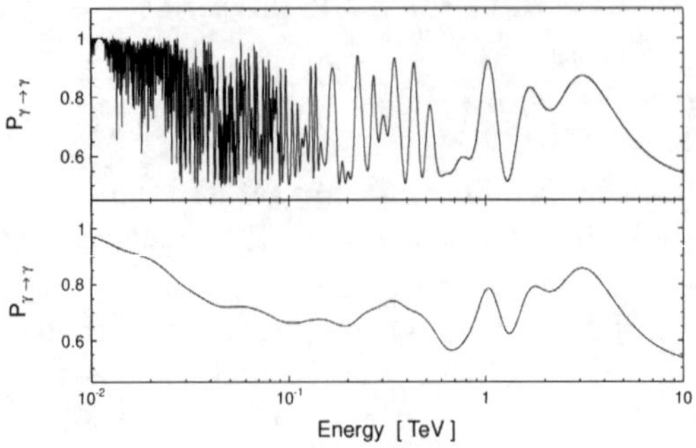

Figure 8. Example of ALP induced irregularity in the TeV range (top panel: Raw signal, bottom panel: Signal smeared with HESS resolution, figure from [38, 39]).

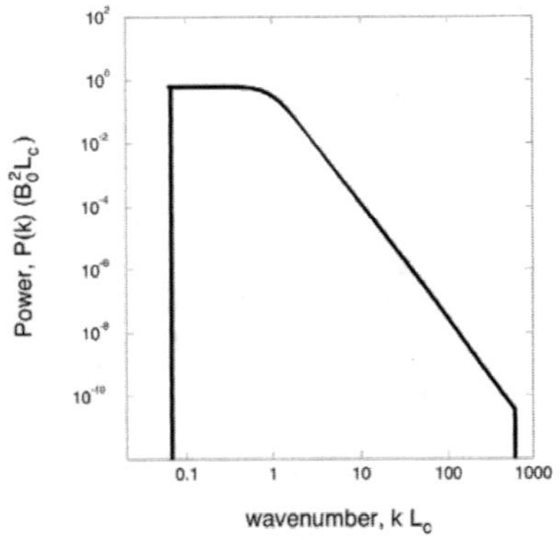

Figure 9. Typical power spectrum used for modeling the magnetic turbulence in galaxy clusters.

cluster magnetic field can be considered. In the first case, as previously, one has to assume optimistic values of the magnetic field strength for the irregularity signal to be significant. A galaxy cluster is observed around the source, but no magnetic field measurements are available. So in the case of the galaxy cluster magnetic field, conservative values for the strength and the coherence length are assumed (1 μG and 10 kpc respectively). As HESS observation ranges from hundreds of GeV to a few TeV, from the expression of the critical energy E_c it is straightforward to see that the typical ALP masses that are probed are of the order of 10^{-8} eV. In [38, 39], it is shown that the observed energy spectrum does not exhibit strong irregularities. Then an estimator of the irregularity is proposed and numerical simulations are used to exclude sets of parameters that lead to significantly too strong irregular behavior. This exclusion has to be done on a statistical basis as each realization of the magnetic field turbulence is different. The results of the analysis are presented in Fig. 10. The method allows to improve the CAST limits in a limited energy range around 20 neV.

Another possibility is to use a source that lies at the center of a well studied galaxy cluster. In that case, the magnetic field properties are derived observationally. This is done by studying the Faraday rotation maps of the polarized radio emission from the cluster (see [40] for a review). These studies allow in principle a determination of the full turbulence power spectrum, yielding the intensity of the magnetic field, its coherence scale and the slope of the turbulence spectrum. A very well studied cluster is Hydra, for which a strong X-ray source is present at the center

(Hydra A) [41] . In [42], X-ray data from the Chandra satellite are analyzed in order to derive constraints on ALP parameters. In the case of X-rays from Hydra, the diagonal terms in the matrix of Eq. 6 can be simplified. Indeed the pair production related opacity is irrelevant in the case of X-rays (so $\tau = 0$), and the trace of the matrix is dominated by the effective photon mass for $m_a \lesssim 10^{-11}$ eV. So the constraints are expected to extend to arbitrarily low ALP masses below that value. In [42], the irregularity is estimated by performing χ^2 tests when deriving the energy spectrum with a forward folding method. ALP parameters yielding a too high level of irregularity compared to the data are excluded. The corresponding exclusion curve is displayed in Fig. 11. It turns out this analysis improves the previous constraints in that mass range from the non-observation of gamma-rays associated with SN 1987 A [43].

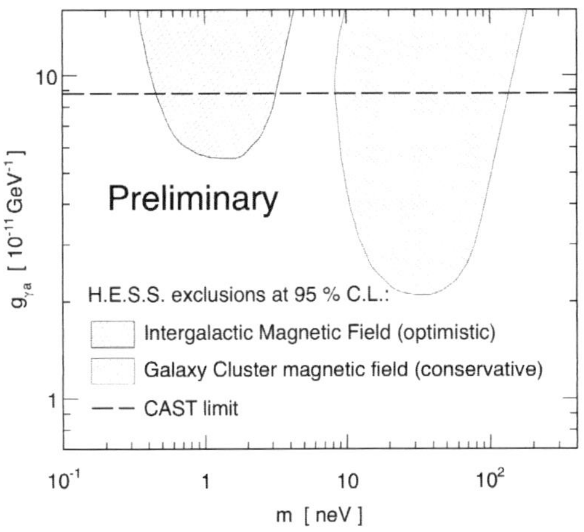

Figure 10. HESS exclusion contours from the analysis of PKS 2155-304 energy spectrum (Figure from [38, 39]).

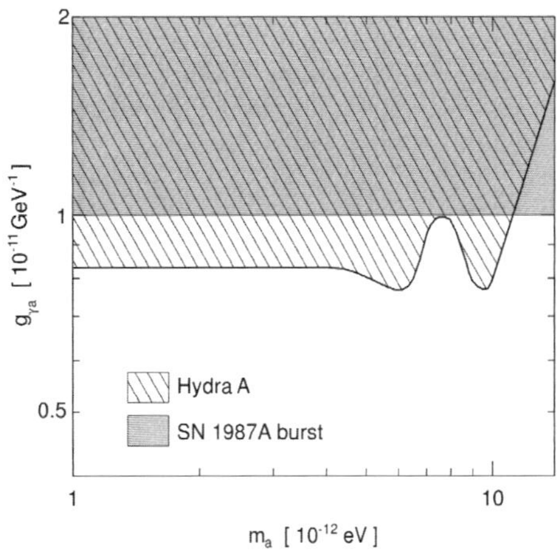

Figure 11. Exclusion contours from the analysis of X-ray data from Hydra A (Figure from [42]).

4. Outlook

The study of the high-energy Universe is potentially a nice way to search for axion-like particles. The problem of the transparency of the Universe to gamma-rays can provide a interesting observable, that requires nevertheless the observation of a large number of TeV sources to be robust. This can be achieved with the next generation of Cherenkov telescopes such as CTA. The photon/ALP mixing in astrophysical sources is intrinsically a stochastic process. That fact makes difficult the use of the transparency observations to derive constraints on the ALP parameters. It is noted however that the turbulence of the astrophysical magnetic fields has the effect of inducing irregularities in the energy spectra of sources. The statistical properties of the induced irregularity can be predicted and are used to set limits on the ALP coupling to photons. Because the method is insensitive to the polarization, these constraints go beyond classic ALPs and apply to both $F\tilde{F}$ and F^2 types of couplings. Two examples of limits are given in the case of TeV and X-ray observations of high-energy emitting sources inside clusters of galaxies.

[1] Beringer J *et al.* (Particle Data Group) 2012 *Phys.Rev.* **D86** 010001
[2] Baker C, Doyle D, Geltenbort P, Green K, van der Grinten M *et al.* 2006 *Phys.Rev.Lett.* **97** 131801 (*Preprint* hep-ex/0602020)
[3] Peccei R D and Quinn H R 1977 *Phys. Rev. Lett.* **38** 1440–1443
[4] Wilczek F 1978 *Phys.Rev.Lett.* **40** 279–282
[5] Weinberg S 1978 *Phys.Rev.Lett.* **40** 223–226
[6] Kim J E 1987 *Phys.Rept.* **150** 1–177
[7] Ringwald A 2012 *Phys.Dark Univ.* **1** 116–135 (*Preprint* 1210.5081)
[8] Arik E *et al.* (CAST) 2009 *JCAP* **2** 008 (*Preprint* 0810.4482)
[9] Aune S *et al.* (CAST) 2011 *Physical Review Letters* **107** 261302
[10] Asztalos S *et al.* (ADMX) 2011 *Nucl.Instrum.Meth.* **A656** 39–44 (*Preprint* 1105.4203)
[11] Ehret K *et al.* (ALPS) 2010 *Phys.Lett.* **B689** 149–155 (*Preprint* 1004.1313)
[12] Svrcek P and Witten E 2006 *JHEP* **0606** 051 (*Preprint* hep-th/0605206)
[13] Arvanitaki A, Dimopoulos S, Dubovsky S, Kaloper N and March-Russell J 2010 *Phys.Rev.* **D81** 123530 (*Preprint* 0905.4720)
[14] Ringwald A 2012 (*Preprint* 1209.2299)
[15] Dole H, Lagache G, Puget J L, Caputi K I, Fernandez-Conde N *et al.* 2006 *Astron.Astrophys.* **451** 417–429 (*Preprint* astro-ph/0603208)
[16] Aharonian F *et al.* (H.E.S.S. Collaboration) 2006 *Nature* **440** 1018–1021 (*Preprint* astro-ph/0508073)
[17] Aliu E *et al.* (MAGIC Collaboration) 2008 *Science* **320** 1752 (*Preprint* 0807.2822)
[18] Becherini Y, Boisson C and Cerruti M (HESS Collaboration) 2012 *AIP Conf.Proc.* **1505** 490–493
[19] Franceschini A, Rodighiero G and Vaccari M 2008 *Astron.Astrophys.* **487** 837 (*Preprint* 0805.1841)
[20] Abramowski A *et al.* (HESS Collaboration) 2012 (*Preprint* 1212.3409)
[21] Wouters D *et al.* (HESS Collaboration) 2012 *AIP Conf.Proc.* **1505** 498–501
[22] HESS Collaboration 2013 *in prep.*
[23] Dominguez A, Primack J, Rosario D, Prada F, Gilmore R *et al.* 2010 (*Preprint* 1007.1459)
[24] Kneiske T M and Dole H 2010 (*Preprint* 1001.2132)
[25] Ackermann M *et al.* (Fermi-LAT Collaboration) 2012 *Science* **338** 1190–1192 (*Preprint* 1211.1671)
[26] Horns D and Meyer M 2012 *JCAP* **1202** 033 (*Preprint* 1201.4711)
[27] Grossman Y, Roy S and Zupan J 2002 *Phys.Lett.* **B543** 23–28 (*Preprint* hep-ph/0204216)
[28] Sanchez-Conde M, Paneque D, Bloom E, Prada F and Dominguez A 2009 *Phys.Rev.* **D79** 123511 (*Preprint* 0905.3270)
[29] Mirizzi A and Montanino D 2009 *JCAP* **0912** 004 (*Preprint* 0911.0015)
[30] Neronov A and Vovk I 2010 *Science* **328** 73–75 (*Preprint* 1006.3504)
[31] Simet M, Hooper D and Serpico P D 2008 *Phys.Rev.* **D77** 063001 (*Preprint* 0712.2825)
[32] Horns D, Maccione L, Meyer M, Mirizzi A, Montanino D *et al.* 2012 *Phys.Rev.* **D86** 075024 (*Preprint* 1207.0776)
[33] Meyer M, Horns D and Raue M 2013 (*Preprint* 1302.1208)
[34] Burrage C, Davis A C and Shaw D J 2009 *Phys.Rev.Lett.* **102** 201101 (*Preprint* 0902.2320)
[35] Pettinari G W and Crittenden R 2010 *Phys.Rev.* **D82** 083502 (*Preprint* 1007.0024)
[36] Wouters D and Brun P 2012 *Phys.Rev.* **D86** 043005 (*Preprint* 1205.6428)
[37] Ostman L and Mortsell E 2005 *JCAP* **0502** 005 (*Preprint* astro-ph/0410501)
[38] Wouters D and Brun P (HESS Collaboration) 2013 (*Preprint* 1304.0700)
[39] HESS collaboration 2013 *in prep.*
[40] Carilli C and Taylor G 2002 *Ann.Rev.Astron.Astrophys.* **40** 319–348 (*Preprint* astro-ph/0110655)
[41] McNamara B, Wise M W, Nulsen P, David L, Sarazin C *et al.* 2000 *Astrophys.J.* **534** L135 (*Preprint* astro-ph/0001402)
[42] Wouters D and Brun P 2013 *accepted in Astrophys. J.* (*Preprint* 1304.0989)
[43] Brockway J W, Carlson E D and Raffelt G G 1996 *Phys.Lett.* **B383** 439–443 (*Preprint* astro-ph/9605197)

Journal of Physics: Conference Series **460** (2013) 012016 doi:10.1088/1742-6596/460/1/012016

Background simulation for the Spherical Proportional Counter and its use for the detection of optical photons

E Bougamont[1], P Colas[1], A Dastgheibi-Fard[1], J Derre[1], I Giomataris[1], G Gerbier[1], M Gros[1], P Magnier[1], X F Navick[1], P Salin[3], I Savvidis[2], G Tsiledakis[1*], J D Vergados[4]
[1]CEA-Saclay, France
[2]University of Thessaloniki, Greece
[3]APC University of Paris, France
[4]University of Ioannina, Greece

Abstract. The recently developed Spherical Proportional Counter [1] allows to instrument large target masses with good energy resolution and sub-keV energy threshold. The moderate cost of this detector, its simplicity and robustness, makes this technology a promising approach for many domains of physics and applications, like dark matter detection and low energy neutrino searches. Detailed Monte Carlo simulations are essential to evaluate the background level expected at the sub-keV energy regime. The simulated background here, it refers to the contribution of the construction material of the detector and the effect of the environmental gamma radiation. This detector due to its spherical shape could be also served as an optical photon detector provided it is equipped with PMTs, for Double Beta decay and Dark Matter searches. All calculations shown here are obtained using the FLUKA Monte Carlo code.

1. Introduction

The development of massive detectors with a low energy threshold and low noise, remains generally a daunting challenge for present-day and future low-background experiments. The search for WIMP (Weakly Interacting Massive Particles) dark matter is under intense development and relies on the detection of low energy (keV scale) recoils produced by the elastic interaction of WIMP's with the nuclei of the detector [2-5]. The question of detecting and exploiting neutrinos from both terrestrial and extra terrestrial sources has become central to physics and astrophysics. Coherent neutrino-nucleus scattering is a famous but as yet untested prediction of the Standard Model [6,7]. Despite having relatively high rates, neutrino-nucleus scattering is difficult to observe because its only signature is a small nuclear recoil of energy ~ keV (for MeV neutrinos). Because the neutrino is light, the nuclear recoil energy is extremely small leading to a signal below threshold for conventional solid or liquid state detectors. Thus, the challenge is to achieve a very low energy threshold (typically below 100 eV). Now thanks to the innovative Spherical Proportional Counter (SPC), very low energy depositions can be measured. Radioactive background studies taking into account the effect of the shield and radioactive contributions from used materials are necessary for the final design and construction for the project that the SPC will be used. In addition, detailed Monte Carlo simulations

[*]georgios.tsiledakis@cea.fr

Journal of Physics: Conference Series **460** (2013) 012016 doi:10.1088/1742-6596/460/1/012016

will be essential to confirm understanding of detector behaviour and response to electron and nuclear recoils, including at the lowest energies and to evaluate the background level expected at the sub-keV energy range relevant for each project.

Figure 1. The rise time (rt) versus the amplitude (in ADC) for a gas mixture of Ar-CH_4 2% where a UV lamp is attached on the sphere.

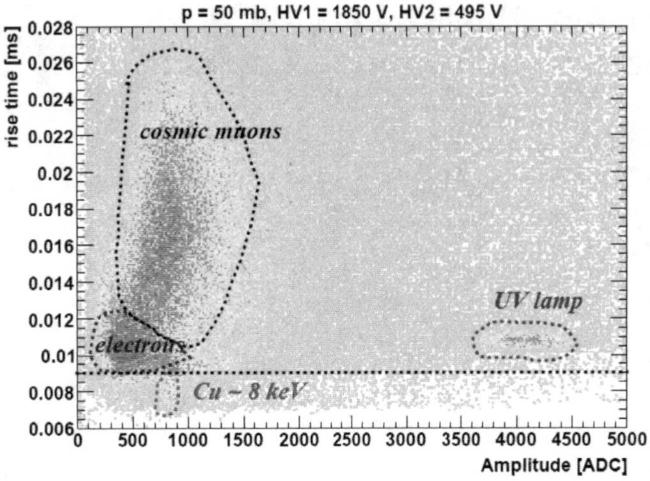

2. Detector description

The detector consists of a large spherical copper vessel 1.3 m in diameter and a small metallic ball 16 mm in diameter located at the center of the drift vessel, which is the proportional counter. The ball is maintained in the center of the sphere by a stainless steel rod and is set at high voltage. A second electrode (umbrella-shaped) that is placed 24 mm away from the ball along the rod, is powered with an independent but lower high voltage, serving as electric field corrector. The detector operates in a seal mode: the spherical vessel is first pumped out and then filled with an appropriate gas at a pressure from few tens of mbar up to 5 bar. Detailed description of the detector, its electronics, its operation and its performance could be found in references [6-8]. Ultra low energy results taken with this counter were presented in reference [1,11], showing an energy threshold as low as 25 eV and a single electron detection sensitivity. The bench mark result is the observation of a well resolved peak at 270 eV due to carbon fluorescence, which is a unique performance for such large massive detector. The main analysis method is to apply fiducial cuts for background rejection. For example, a run is performed with the present detector installed at ground (Saclay), using a gas filling of Argon with 2% admixture of CH_4 and having a UV flash lamp installed in one of the sphere openings [12]. The scatter plot of the rise time (rt) versus the amplitude of the signal is shown in Figure 1. The rise time of the signal actually provides the depth of the ionized electrons produced in the gas. By applying a cut at $rt < 0.009$ ms, we keep only volume events achieving a spectacular background reduction. For underground tests another similar detector of 1.3 m in diameter has been installed in the LSM laboratory in Modane (Frejus) under 1700 m of rock (4800 m water equivalent) providing protection from cosmic rays. Comparisons between measurements in the sub-keV energy region taken with both detectors need to carried out in order to understand the background level and optimize the detector in terms of sensitivity and noise background. In this work we will focus our studies to simulate the background.

3. Background simulations

3.1. FLUKA Monte Carlo code

FLUKA is a fully integrated particle physics Monte Carlo simulation package [13]. It has many applications in high energy experimental physics and engineering, shielding, detector and telescope design, cosmic ray studies, dosimetry, medical physics and radio-biology. For our simulations presented here, the EM-CASCA defaults card is used which provides the defaults for pure electromagnetic (EM) cascades. Its main features are: - Electromagnetic interactions are on - Rayleigh scattering and inelastic form factor corrections to Compton scattering and Compton profiles activated - Detailed photoelectric edge treatment and fluorescence photons activated - Restricted ionisation fluctuations for EM - Both explicit and continuous heavy particle bremsstrahlung and pair production inhibited. The present lowest transport limit for electrons and photons is 1 keV.

3.2. *Effect of cobalt-60*

The spherical vessel of the SPC is made by copper which is not particularly pure and clean. Among the probable contaminants, the most potentially dangerous is the radioactive isotope cobalt-60 which has half-life 5.27 years and produces 2 gammas rays of 1.17 and 1.33 MeV. These gammas can interact with the copper itself through Compton and create a gamma background inside the gas. This effect has been simulated using the FLUKA MC as follows: - the geometry of the SPC is implemented with its actual dimension as a spherical copper vessel that surrounds a gas of Ar-CH_4 (2%) at pressure 50 mb. - two monochromatic photons of 1.33 and 1.17 MeV originating in a random position in the copper vessel are emitted according to an isotropic angular distribution (R sampling proportional to R^2). – two independent runs of 3M events for each photon energy are produced – the results are combined and the energy deposition inside the gas is calculated on event-by-event basis. To separate electron and photon contributions, since no automatic distinction is possible at scoring time (because all energy deposition is done by electrons, even when the primary is photon), we could selectively kill electrons (or photons) at the boundary between copper and gas. Figure 2 shows the energy deposition inside the gas from photons (the electrons are killed in the boundary) originating from cobalt-60. The two peaks in the left plot indicate the two single gamma lines around 8 keV from the induced fluorescence at the copper vessel. Adjusting a detector resolution 10% at FWHM we obtain the right plot.

Figure 2. The energy deposition obtained by FLUKA simulation inside the gas of the SPC when only photons, originating from cobalt-60, are entering from the Cu vessel inside the gas

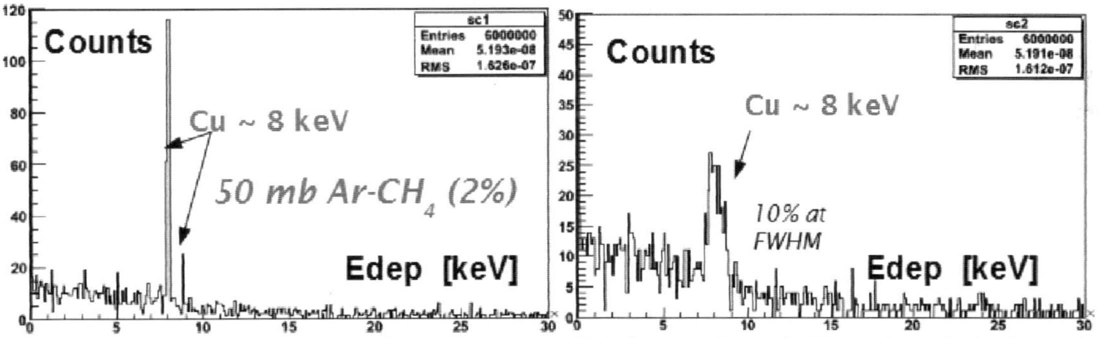

The Monte Carlo gives ~ 190 hits in copper out of 3M decays of cobalt. Thus, there is 1 photon in copper per ~ 16000 decays of cobalt-60. The measured data at LSM give a rate in copper ~ 1Hz. Provided that the measured rate is fully attributed to the cobalt-60, the expected radioactivity should be ~ 16000 kBq. But the measured radioactivity at the LSM on the SPC copper gives an upper limit for the cobalt-60 < 1.3 mBq/kg [14] or A = 1.3mBq*200 kg = 0.26 Bq ➔ f_{Cu}(data) ~ 1.6E-5 Hz.
As a result, the contribution from cobalt-60 at the measured rate in copper is negligible. So the environmental gammas should induce the main signal on copper at LSM as it is indicated at the next session.

3.3. *Estimation of the photon background at LSM*

After showing that cobalt-60 cannot explain the measured ~ 1 Hz rate in copper, the next step is to see what the effect of the environmental gammas is. Figure 3 shows a table with photon fluxes of the most prominent lines at LSM at various locations. There are uncertainties on these fluxes roughly a factor 2, depending on the location of measurement. With a simulation it could be possible to predict the energy spectrum observed in the SPC installed at LSM (Modane) and to provide a first hint to simulate real background spectra.

Figure 3. The photon flux at LSM at various locations

Extract of report (2004) by R Gurriaran about measurement of photon flux in LSM at various locations (xxl, nemo ,...) for most prominent lines.

Unit = 10^{-3} photon/cm^2/s

tableau1: mesure du flux exprimé en ph.s^{-1}.cm^{2} (.10^{-3})

radionucléide	^{40}K		^{214}Pb		^{214}Bi		^{212}Pb		^{208}Tl		^{208}Tl		^{137}Cs		^{60}Co		^{60}Co	
énergie (keV)	1460,8		351.9		609,3		238,6		583,2		2614,5		661,7		1173,2		1332,5	
		+/-		+/-		+/-		+/-		+/-		+/-		+/-		+/-		+/-
xxl (blindage)	76	2,2	25,5	0,4	40,5	0,5	8,18	0,3	8,5	0,3	26,1	1	1,44	0,2				
nemo (blindage)	85,5	2,5	21,2	0,8	35,3	2,8	6,32	0,6	5,67	0,7	27,1	1,7	7,14	0,7				
radon	125	4,4	42,6	1,7	66,6	2,4	13,7	1,1	14,3	1,3	42,1	6,9						
jasmin	127	5,1	36,3	1,8	64,3	6,3	11,6	1,2	12,6	1,4	40,7	3,4	2,26	0,9	12,9	1,8	13,2	1,8
tgv	141	2,8	25	0,8	41,4	1,2	7,35	0,6	9,23	0,7	29,1	1,6	5,62	0,6				
edelweiss	207	3,1	47,1	1,1	72,2	0,6	17,3	0,3	18,1	1	51,1	1,9	1,07	0,2				
bureau (poiser)	434	7	51,5	0,6	81,7	0,9	50,8	2,7	51	0,7	126	4	3,73	0,4				

Figure 4. The energy deposition obtained by FLUKA simulation inside the gas of the SPC for various pressures, originating from the seven most prominent gamma lines at the edelweiss location

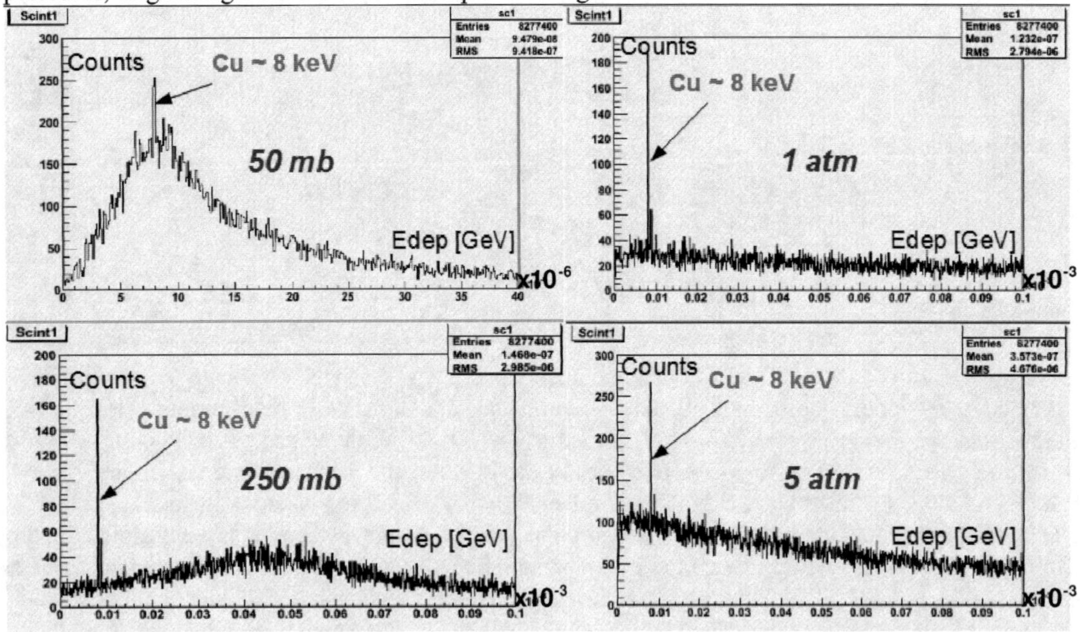

As an input for the simulation, the values of the fluxes at the edelweiss location (next to where the SPC was placed) are used. Seven FLUKA analogue runs took place (one for each of the seven photon lines) and then the results were merged. The methodology is similar with the one in the previous subsection with the difference that each of these gammas is coming from an extended source shaped as a spherical shell with radius the external one of the SPC copper vessel and 4π isotropic direction. In all cases a mixture of Ar-CH_4 (2%) is used at pressures of 50 mb, 250 mb, 1 atm and 5 atm. The energy deposition spectra, obtained by FLUKA simulation, inside the gas of the SPC for various pressures, originating from the seven most prominent gamma lines at the edelweiss location are shown in Figure 4. At 50 mb a wide distribution mostly from electrons is shown and is peaked around 8 keV. By increasing the pressure this distribution is moving towards higher energies. By killing the electrons at the boundary between copper and gas we end up in the energy deposition spectra in Figure 5 (this can be done with the real data thanks to the analysis method by applying fiducial cuts using the rise time measure). At higher pressures, the rather steady background hit rate which is actually due to Compton electrons inside the gas, becomes more dominant.

Figure 5. The energy deposition obtained by FLUKA simulation inside the gas of the SPC for various pressures, originating from the seven most prominent gamma lines at the edelweiss location when only photons are entering from the Cu vessel inside the gas.

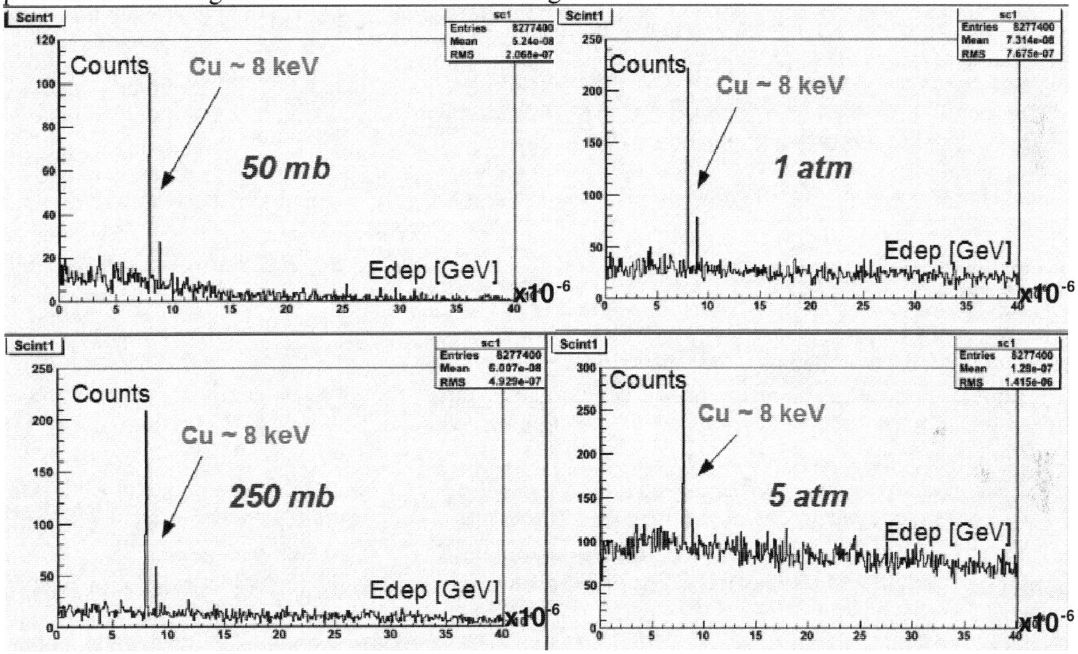

In LSM, at pressure 50 mb, the measured rate on copper is ~ 1 Hz. FLUKA MC gives ~ 200 hits in copper out of 8.3M photons (=Nevents). The measured total flux at the edelweiss location is: 413.9E-3 g/cm^2/s, the surface of the SPC: $4\pi R^2$=53093 cm^2, 1 sec→ 53093*413.9E-3=21974 gammas. Thus, the rate in copper is: f_{Cu} = 200*21974/8.3M = 0.53 Hz at 4π direction. Adding a factor 2 due to the preferential direction of the gammas towards the sphere according to the definition of the flux, we conclude that the simulated rate in copper is ~ 1 Hz which comes into agreement with the data.

The scatter plot of the energy deposition at gas pressure of 1 atm, versus the radius of the SPC is shown in Figure 6. A photon with energy of 1461 keV hits the SPC copper vessel having a 4π isotropic direction and only short tracks are kept requiring 1 hit inside a cell of 1 cm side (the long tracks have large rise time and can be excluded with the data analysis method). It is obvious that most of the local energy depositions are at the periphery of the SPC gaseous volume and by performing fiducial cuts we could decrease the background.

Figure 6. The energy deposition obtained by FLUKA simulation inside the gas versus the radius of the SPC

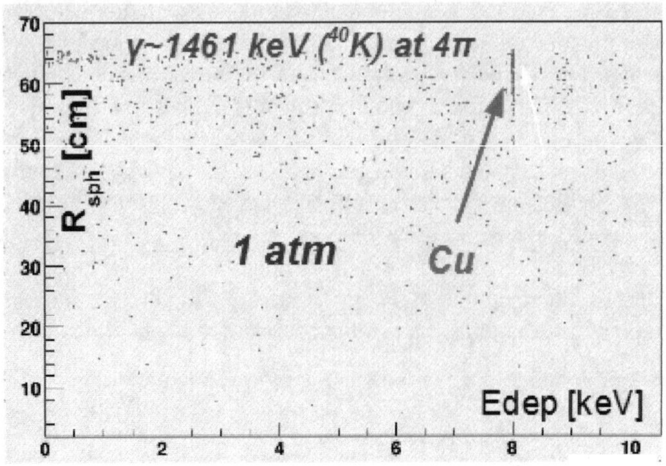

4.The use of SPC as an optical photon detector

In the WIMP's search, the XENON project is one of several direct detection experiments worldwide, using a noble liquid as target and detector medium. The XENON100 detector is a double-phase (liquid-gas) time projection chamber (TPC). The ratio S2/S1 (where S1 is the primary scintillation light and S2 the secondary) allows event discrimination between nuclear recoils (WIMPs, neutrons) and electron recoils (γ, β) [15].

A spherical proportional counter (SPC) made by Al (served as UV light reflector) and filled with high pressure noble gas, can be applied for the detection of optical photons. Simulations are performed using the FLUKA MC for the estimation of the light collection efficiency (LCE) of the S1 on the installed photomultipliers (PMTs).

Optical photons can be generated and propagated in FLUKA [13] and can be treated according the laws of geometrical optics. Therefore, they can be reflected and refracted at boundaries between different materials, absorbed in flight or elastically scattered (Rayleigh scattering).

In the simulation below, 6 PMTs of 12.5 cm in diameter are placed on the SPC (top left plot in Figure 7) and the LCE on them is estimated considering that the surrounding vessel has a reflectivity of 92%. At first approach there is no absorption in flight and no Rayleigh scattering. 2k optical photon sources are generated randomly inside the SPC and each of them emits at 4π direction 1k optical photons.

Two types of reflection are used: the specular (mirror-like) and the diffuse reflection. Without reflection, the LCE of 6 PMTs is: $6*A_{PMT}/A_{SPH} \sim 1.4\%$. Thus, for a SPC with radius of 65 cm and 6 PMTs, a diffuse reflection of 92% and 99% improves the LCE by ~ 10 times (LCE = 15%) and ~ 40 times (LCE = 58%) respectively. In the case of 99% diffuse reflection, if there is an absorption length of 50 m and the Rayleigh scatter length is of 24 m, the LCE drops to ~ 38%.

Figure 7 summarises these results with the addition of the use of one single PMT placed on the centre of the SPC.

A spherical proportional counter with the present dimensions and with just 6 PMTs installed around, presents a S1 LCE large enough to provide event trigger for a 100 keV gamma, enough for DBD searches and good sensitivity for O(keV) events in the field of DM searches . Such an event trigger can provide radial-position information for fiducial volume selection.

Figure 7. The light collection efficiency in the SPC, obtained by FLUKA simulation, for specular and diffuse reflection and the particle tracklength density spectra (fluence) expressed in particles/cm^2.

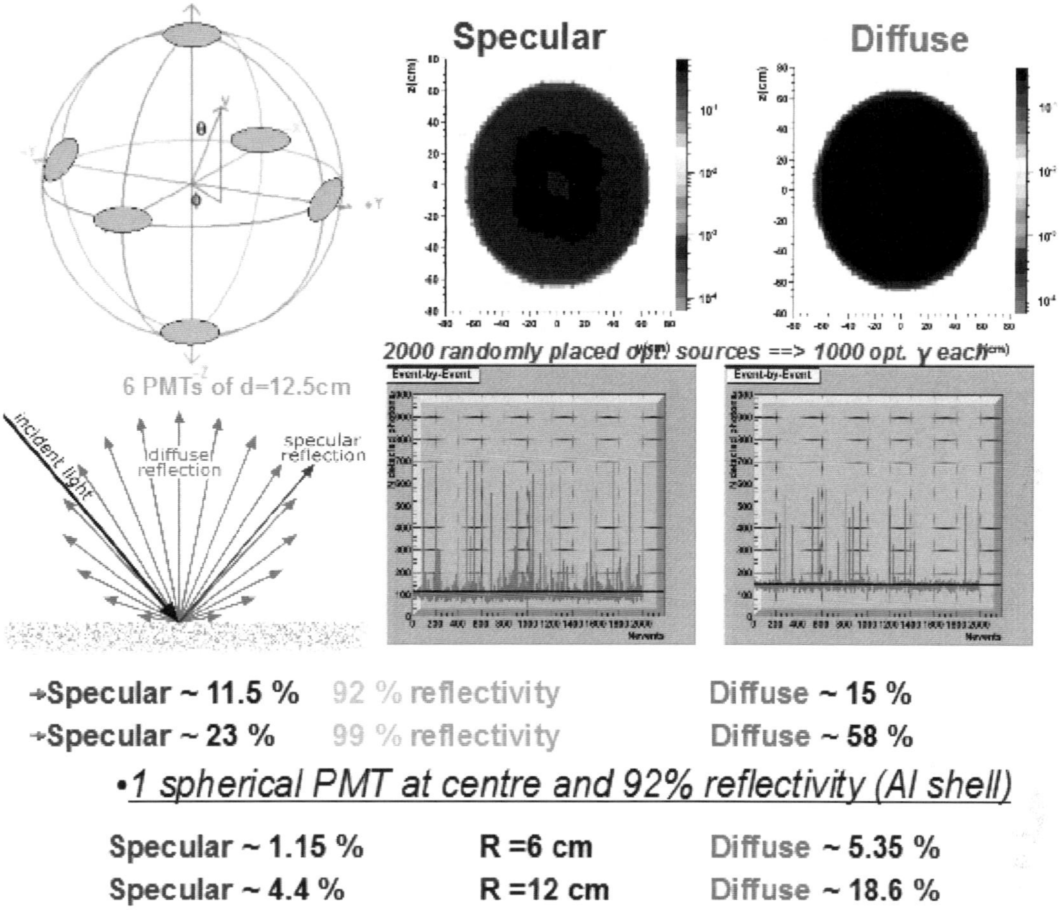

5. Conclusions

A new detector has been developed that combines large mass and low sub-keV energy threshold with the capability of sensing even single electrons. Detailed simulations are essential to evaluate the background level expected at the sub-keV energy range relevant for each project that this type of detector can be used. Its moderate cost, simplicity, robustness and versatility since it can be served as an optical photon detector, make this technology a promising approach opening a new window in dark matter searches.

REFERENCES

[1] E. Bougamond et al., arXiv: 1010.4132 [physics.ins-det], to be published at JMP (2010).
[2] K. Nakamura et al. (Particle Data Group), J. Phys. G 37: 075021 (2010).
[3] I. Giomataris and J.D. Vergados, Nucl. Instrum. Meth. A530: 330-358 (2004).
[4] J.I. Collar, Y. Giomataris, Nucl. Instrum. Meth. A471: 254-259 (2000).

[5] R. Bernabei et al., Eur. Phys. J. C67: 39-49 (2010).
[6] D.Z. Freedman et al., Ann. Rev, Nucl. Sci. 27: 167 (1977).
[7] A. Drukier and L. Stodolsky, Phys. Rev. D30: 2295 (1984).
[8] S. Aune et al., AIP Conf. Proc. 785: 110-118 (2005).
[9] I. Giomataris et al., Nucl. Phys. Proc. Suppl. 150: 208-213 (2006).
[10] I. Giomataris et al., JINST 3: P090007 (2008).
[11] E. Bougamond et al., J.Phys. Conf. Ser. 309 012023 (2011).
[12] E. Bougamond et al., Conf.Proc: www-library.desy.de/preparch/desy/proc/proc11-03.pdf (2012).
[13] "The FLUKA code: Description and benchmarking" G. Battistoni, S. Muraro, P.R. Sala, F. Cerutti, A. Ferrari, S. Roesler, A. Fasso`, J. Ranft, Proceedings of the Hadronic Shower Simulation Workshop 2006, Fermilab 6--8 September 2006, M.Albrow, R. Raja eds., AIP Conference Proceeding 896, 31-49, (2007)
"FLUKA: a multi-particle transport code" A. Fasso`, A. Ferrari, J. Ranft, and P.R. Sala, CERN-2005-10 (2005), INFN/TC_05/11, SLAC-R-773,
www.fluka.org
[14] Private communication with P. Loaiza from LSM laboratory
[15] E. Aprile (XENON coll.) Journal of Physics: Conference Series 203 (2010) 012005

Low Energy Investigations at Kamioka Observatory

Hiroyuki Sekiya

Kamioka Observatory, Institute for Cosmic Ray Research, University of Tokyo, 456 Higashi-Mozumi, Kamioka, Hida, Gifu, 506-1205 Japan

E-mail: sekiya@icrr.u-tokyo.ac.jp

Abstract. At Kamioka Observatory many activities for low energy rare event search are ongoing. Super-Kamiokande(SK), the largest water Cherenkov neutrino detector, currently continues data taking as the fourth phase of the experiment (SK-IV). In SK-IV, we have upgraded the water purification system and tuned water flow in the SK tank. Consequently the background level was lowered significantly. This allowed SK-IV to derive solar neutrino results down to 3.5 MeV energy region. With these data, neutrino oscillation parameters are updated from global fit; $\Delta m^2_{12} = 7.44^{+0.2}_{-0.19} \times 10^{-5}\mathrm{eV}^2$, $\sin^2\theta_{12} = 0.304 \pm 0.013$, $\sin^2\theta_{13} = 0.030^{+0.017}_{-0.015}$. NEWAGE, the directional sensitive dark matter search experiment, is currently operated as "NEWAGE-0.3a" which is a $0.20 \times 0.25 \times 0.31$ m^3 micro-TPC filled with CF4 gas at 152 Torr. Recently we have developed "NEWAGE-0.3b". It was succeeded to lower the operation pressure down to 76 Torr and the threshold down to 50 keV (F recoils). XMASS experiment is looking for scintillation signals from dark matter interaction in 1 ton of liquid xenon. It was designed utilizing its self-shielding capability with fiducial volume confinement. However, we could lower the analysis threshold down to 0.3 keVee using whole volume of the detector. In February 2012, low threshold and very large exposure data (5591 kg·days) were collected. With these data, we have excluded some part of the parameter spaces claimed by DAMA/LIBRA and CoGeNT experiments.

1. Kamioka Observatory

Kamioka Observatory, Institute for Cosmic Ray Research, University of Tokyo, is located in Kamioka-town in the northern part of the Gifu prefecture in Japan. The location of the laboratories is under the peak of Mt. Ikenoyama providing 2,700 meters water equivalent (or 1000 m of rock) overburden. This observatory was established in 1995 to push forward with the Super-Kamiokande experiment and has become a world frontier center of neutrino physics. Many studies complementary to the flagship SK are also ongoing as illustrated in Figure 1. Experiments to search for dark matter and double beta decay are very active now. Further a new 3 km gravitational wave telescope is being constructed and the next generation flagship experiment: Hyper-Kamiokande is seriously discussed in Japanese physics community.

2. Super-Kamiokande

Super-Kamiokande (SK) is a water Cherenkov detector containing 50,000 tons of pure water viewed by 11,129 (inner detector: 32 kton) and 1,885 (outer detector: 18 kton) photomultiplier tubes (PMTs). A broad range of energies is studied with the SK detector, from a few MeV up to O(TeV). In the low energy region below ~20 MeV, solar neutrino interactions are detected by neutrino-electron scattering. At higher energies, atmospheric neutrino interactions

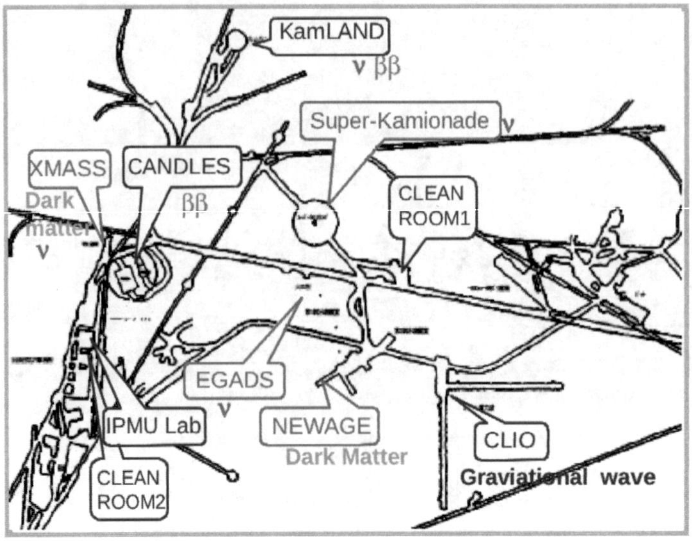

Figure 1. Kamioka underground map of experiments.

covering 5 orders of magnitude are detectable by neutral- and charged-current neutrino-nucleon interactions.

In this presentation, we focus on the lower energy physics, solar neutrino. SK detects solar neutrinos through neutrino-electron elastic scattering, where the energy, direction, and time of the recoil electron are measured. Due to its large (22.5 kton) fiducial mass SK makes precise measurements of ^{8}B solar neutrinos, including precision information on their energy spectrum and its time variation.

Since SK-III started, we have made many efforts to reduce backgrounds. The most serious background comes from the beta decay of ^{214}Bi, which is produced in the decays of radon in the air and detector materials as well as from radium in the water. In order to reduce the ^{214}Bi background, the SK water system has been upgraded. First, the cooling power of the heat exchanger for the supply water was increased so as not to induce convection in the tank, which transports radon near the PMTs into the fiducial volume. Second, new membrane degasifier units were added to increase the efficiency of radon removal from the supply water. Third, the water flow in the detector was precisely investigated and optimized to reduce the background contamination in the fiducial volume as much as possible. During the SK-IV period we have introduced a precise temperature control system for the inlet water to further reduce convection in the water.

As a result of these improvements, the background rate in the lower energy region in SK-IV has been reduced by a factor of more than three compared to the SK-I period. Further, it has allowed a lower analysis energy threshold. Until the end of March 2012, 1069.3 days of SK-IV solar neutrino data for analysis was taken. A clear solar neutrino signal in the 3.5-4.0 MeV kinetic energy region was observed at more than $7\,\sigma$. In addition, we developed a new analysis method for these low energy regions based on the amount of multiple Coulomb scattering of electrons using the PMT hit pattern of the Cherenkov cone. Lower energy electron hit patterns (i.e. the ^{214}Bi background) are more isotropic than those of the higher energy solar neutrino recoil electrons. This analysis improves the statistical uncertainty of the number of signal events by about 10 %. We use this new method for recoil energies below 7.5 MeV.

Figure 2 shows SK-I to SK-IV combined energy spectrum with expectations from the solar

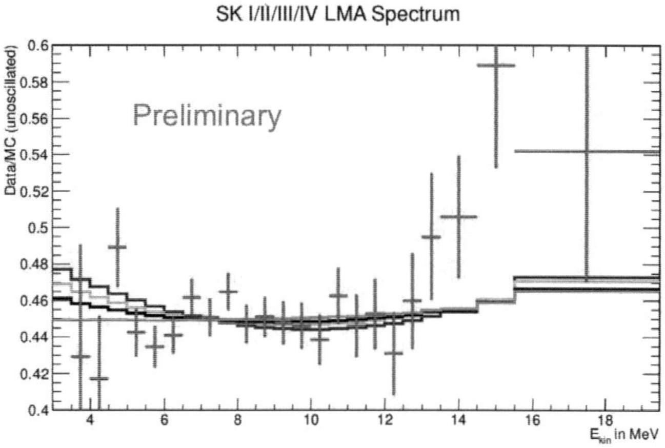

Figure 2. SK-I to SK-IV combined solar neutrino energy spectrum. Each point shows the ratio of the data to the expected flux using an unoscillated ^{8}B solar neutrino spectrum. Predictions for (1) $\sin^2\theta_{12} = 0.304$ and $\Delta m^2_{21} = 7.4 \times 10^{-5}$eV^2(blue), (2) $\sin^2\theta_{12} = 0.314$ and $\Delta m^2_{21} = 4.8 \times 10^{-5}$eV^2(light blue), (3) flat probability (black), and (4) flat probability and $d\sigma/dE$ shape for pure $\nu_e + e$ scattering (gold) are also shown.

global and solar+KamLAND as well as flat reduction of the neutrino spectrum. The vertical axis shows the ratio of the observed energy spectrum to the expectation from the unoscillated MC simulation assuming a ^{8}B flux of 5.25×10^6 cm^{-2}s^{-1}. The combined energy spectrum is consistent with the flat prediction, but the level of favoring flat over the upturn is $1.1{\sim}1.9\sigma$ level.

Concerning differences in the day and night fluxes the expected flux asymmetry, defined as $A_{DN} = (day - night)/\frac{1}{2}(day + night)$, is about -2% based on current understanding of neutrino oscillation parameters. Although this is not a large effect, long term observations by SK enable discussion of a finite value of the day-night asymmetry. The A_{DN} value using the combined SK-I to SK-IV data is $-2.8 \pm 1.1 \pm 0.5\%$, which is a 2.3σ difference from zero. Figure.3 shows A_{DN} as a function of Δm^2 together with the expectation. The observed A_{DN} is consistent with the expectation using the best fit Δm^2 from both KamLAND and the global solar analysis.

We performed a global solar neutrino oscillation analysis including all SK data (SK-I [1],SK-II [2], SK-III [3], and SK-IV) as well as the most recent results from SNO [4], the radiochemical experiments [5, 6] and the latest ^{7}Be flux measurement from Borexino [7]. This analysis was then compared and combined with the reactor neutrino results from KamLAND [8]. The obtained range of $\sin^2\theta_{13}$ from the solar global analysis is $0.014^{+0.027}_{-0.021}$, while the value of the KamLAND reactor analysis is $0.031^{+0.038}_{-0.036}$. In the combined fit $\sin^2\theta_{13}$ was found to be $0.030^{+0.017}_{-0.015}$, which is a roughly 2 σ hint that θ_{13} is different from zero. This has been already discussed prior to 2010 [3]. After 2011, the T2K and reactor experiments presented indications and later evidence for a finite θ_{13}. The combination of their measurements yields $\sin^2\theta_{13} = 0.025^{+0.003}_{-0.004}$ and the result of the combined analysis of the global solar and KamLAND reactor data is consistent with this value.

Assuming $\sin^2\theta_{13}$ is fixed at 0.025, the obtained parameters from the global solar analysis are $\Delta m^2_{21} = (4.86^{+1.44}_{-0.52}) \times 10^{-5}$eV^2 and $\sin^2\theta_{12} = 0.310^{+0.014}_{-0.015}$. Comparing these values with those from KamLAND, ($\Delta m^2_{21} = (7.49^{+0.20}_{-0.19}) \times 10^{-5}$eV^2 and $\sin^2\theta_{12} = 0.309^{+0.039}_{-0.029}$), there is a 1.8 σ level tension in the Δm^2_{21} results. Combining the global solar data with KamLAND, the oscillation parameters become $\Delta m^2_{21} = (7.44^{+0.20}_{-0.19}) \times 10^{-5}$eV^2 and $\sin^2\theta_{12} = 0.304 \pm 0.013$.

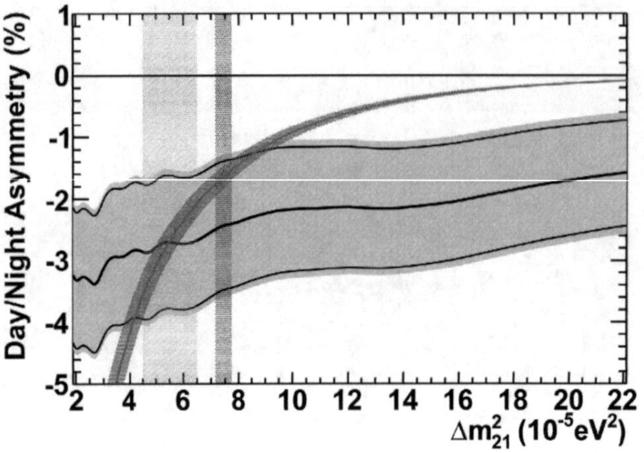

Figure 3. The day-night asymmetry as a function of Δm^2. A thick black line shows the result of the SK fit and the surrounding gray band indicates statistical and systematic uncertainty (thin black lines for statistical error only). The red curve shows the expectation assuming $\sin^2(\theta_{12}) = 0.314$. The light blue and light green vertical bands show the allowed range of Δm^2 from the KamLAND reactor and solar global, respectively.

3. NEWAGE

Among the most plausible dark matter candidates, Weakly Interacting Massive Particles (WIMPs), which are expected to couple to ordinary matter primarily through the weak force, can be detected directly through observation of nuclear recoils produced in their elastic scattering interactions with detector nuclei [9]. The most convincing signature of the WIMPs appears in the directions of nuclear recoils. It is provided by the earth's velocity through the galactic halo (~230 km/s) and the distribution of the nuclear recoil direction shows a large asymmetry. Hence, detectors sensitive to the direction of the recoil nucleus would have a great potential to identify WIMPs. Time Projection Chambers (TPCs) with fine spacial resolutions are among such devices, and we are developing a micro-TPC [10, 11], which can detect three-dimensional fine tracks of charged particles. Since the energy deposits of WIMPs to nuclei are only a few tens of keV and the range of nuclei is limited, the micro-TPC should be operated at low pressures. We are interested in operating the micro-TPC with CF_4 because ^{19}F has a special sensitivity to SD interactions for its unique spin structure.

We performed our first dark matter search in 2010 at Kamioka laboratory using NEWAGE-0.3a detector [12]. The exposure was 0.524 kg·days. The gas pressure was 152 Torr and the energy threshold was 100 keV. Then, second search was conducted from January 2012 to May 2012. This exposure was 0.140 kg·days with the same gas pressure as before and the energy threshold was lowered to 50 keV. Figure 4 shows the energy spectra and exclusion limits obtained through these searches. We set a limit of 800 pb for 100 GeV WIMPs, however it is indispensable to lower the threshold and background. In this presentation, we report on the study of lowering the energy threshold via operating the detector in lower pressure.

The low pressure gas study was performed using NEWAGE-0.3b detector in the surface laboratory. A schematic drawing of NEWAGE-0.3b is shown in Figure 5 The size of the μ-PIC (No.SN070309-3) is 30 cm×30 cm, and the number of pixels is 768×768. We supply a positive high voltage to the anode electrodes of the μ-PIC (= V_{ANODE}). A large area GEM, consisting of a 50 mm-thick polyimide layer sandwiched between 5 mm-thick copper electrodes is set at 5 mm

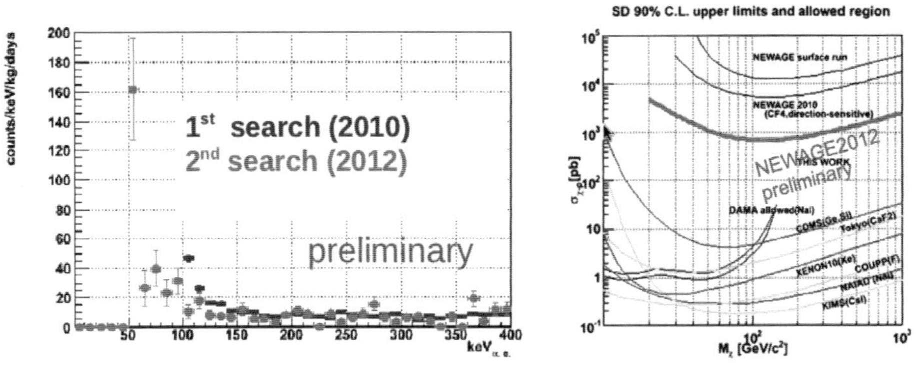

Figure 4. Obtained energy spectra and exclusion limits by NEWAGE-0.3a detector at Kamioka.

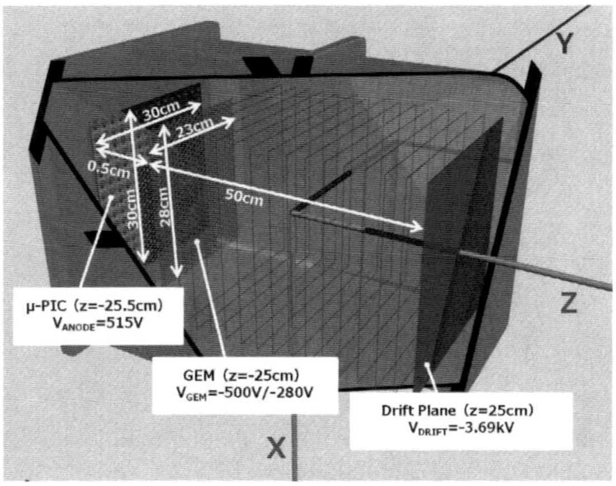

Figure 5. Schematic drawing of NEWAGE-0.3b detector. Induction gap, the distance between μ-PIC and GEM, is 5 mm.

above the μ-PIC to obtain sufficient gain. The size of the GEM is 23 cm$\times$28 cm, and this GEM is segmented into 8 parts to reduce the risk by discharges. We supply negative high voltages at the top (gas-volume side: V_{GEMtop}) and bottom (μ-PIC side: $V_{\mathrm{GEMbottom}}$) of the GEM. The drift length is 50 cm and the detection volume is surrounded by a drift cage to form the electric field. The size of the drift cage is $36.3 \times 36.3 \times 51$ cm. Copper electrodes are formed on the inner surface of the drift cage by every 1 cm. We supply a positive high voltage ($= V_{\mathrm{DRIFT}}$) to the drift plane. These detector components are set in a chamber made of SUS304 stainless steel whose inner surface was electro-polished to reduce out-gassing. The chamber was filled with CF_4 gas at 76 Torr. A dedicated electronics system was used for the measurement. Details of the electronics are described in [12] and the references therein. Three-dimensional tracks were detected by the discriminated hit signals of 768 anode strips and 768 cathode strips and the anode-cathode coincidence-timing. Energy of each track was determined by the summed waveform.

As a beginning, the gain dependence on the bias voltages was measured to optimize operation parameters for CF_4 at 76 Torr. We had operated the previous NEWAGE-0.3a at a gas gain of 700 with 152 Torr gas. Accordingly, the optimum gas gain should be 1400 at 76 Torr, because the primary electron density became 1/2 of the previous detector. In Figure 6, the obtained gas

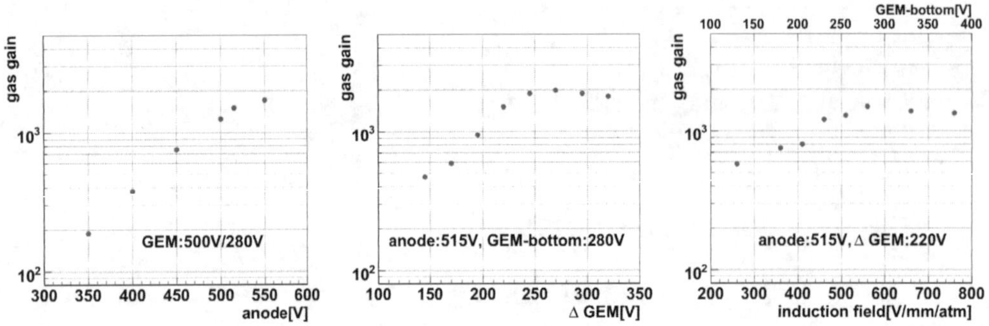

Figure 6. Measured gas gain curves against V_{ANODE}, ΔV_{GEM} and $V_{\mathrm{INDUCTION}}$ (or $V_{\mathrm{GEMbottom}}$).

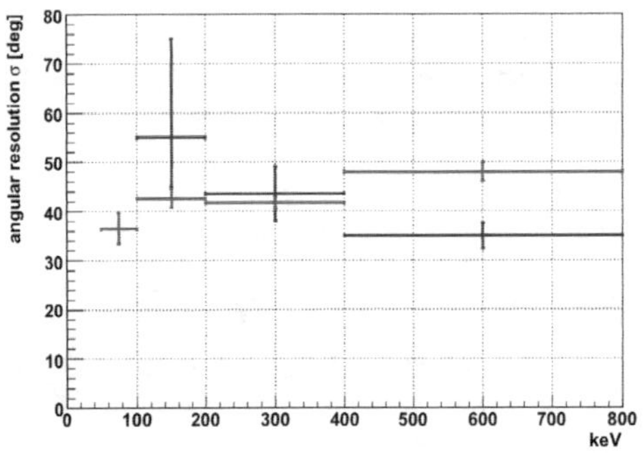

Figure 7. Energy dependence of measured angular resolutions. Red and blue histograms show the results of 76 Torr and those of 152 Torr, respectively.

gain against V_{ANODE}, $\Delta V_{\mathrm{GEM}} = V_{\mathrm{GEMtop}} - V_{\mathrm{GEMbottom}}$, and the induction field ($V_{\mathrm{INDUCTION}}$). The other two parameters are fixed in each plot. The gas gain increased with V_{ANODE} until the voltage was limited by discharged. As for ΔV_{GEM} and $V_{\mathrm{INDUCTION}}$, gain curves saturate below 2000. This gain suppression is due to the operation in low gas pressure. The mean free path of an electron in 76 Torr of CF_4 gas is several mm, and this length is close to the thickness of the GEM (50 mm), therefore the gas amplification at the GEM is to be suppressed. Although it turned out that the gain is saturated, we have achieved the required gas gain of 1400. The optimized parameters are $V_{\mathrm{ANODE}} = 515\,\mathrm{V}$, $V_{\mathrm{GEMtop}} = 500\,\mathrm{V}$, $V_{\mathrm{GEMbottom}} = 280\,\mathrm{V}$ and $V_{\mathrm{DRIFT}} = 3.69\,\mathrm{kV}$. Then, the angular resolution of NEWAGE-0.3b was evaluated using nuclear tracks by neutrons from a ^{252}Cf source. We determined the angular resolution by comparison of measured and simulated distributions of the recoil angle. The results are shown in Figure 7.

We plan to move the NEWAGE-0.3b detector to Kamioka underground in 2013. In order to operate it with lower pressure gas, we need an even higher gas gain. As shown in above the gain is saturated at around 2000 in the current set up, therefore we have started to test a new GEM which has a thickness of 100 mm instead of 50 mm. We are also improving the data acquisition system (DAQ) for recording the time-over-threshold (TOT) of the μ-PIC signals in order that we can measure the energy deposition at each strip.

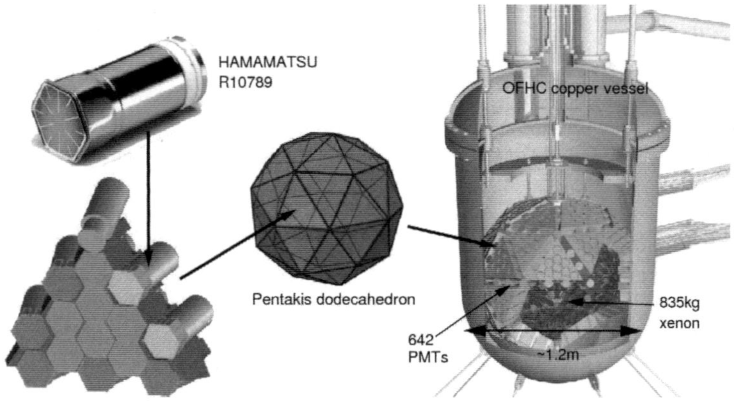

Figure 8. Structure of the XMASS detector.

4. XMASS

Over the past 10 years, some experiments indicate a possible WIMP signal [13, 14, 15] with a lighter mass of $\sim 10\,\mathrm{GeV}$ and with a spin-independent cross section of the order of $\sim 10^{-40}\mathrm{cm}^2$. These positive signals have come predominantly from experiments without the capability to discriminate between electromagnetic and nuclear recoils, while other experiments that have this ability have excluded light WIMPs at these cross-sections [16, 17, 18]. In this presentation, a search for light WIMPs using a 5591.4 kg·day exposure of the XMASS experiment without nuclear recoil discrimination is presented.

XMASS is a single phase liquid xenon scintillator detector containing 1050 kg of Xe in an OFHC copper vessel. As shown in Figure. 8 xenon scintillation light is collected by 630 hexagonal and 12 cylindrical inward-pointing Hamamatsu R10789 series photomultiplier tubes (PMTs) arranged on an 80 cm diameter pentakis-dodecahedron support structure within the vessel to give a total photocathode coverage of 62.4 %. These PMTs view an active target region containing 835 kg of liquid xenon. In order to monitor the PMT stability and measure the trigger efficiency, eight blue LEDs with Teflon diffusers are mounted to the support structure. There are six LEDs arranged along the equator and one each at the top and the bottom of the pentakis-dodecahedron. To shield the scintillator volume from external gammas, neutrons, and muon-induced backgrounds, the copper vessel is placed at the center of a $\phi\,10\,\mathrm{m} \times 11\,\mathrm{m}$ cylindrical tank filled with pure water. This volume is viewed by 72 Hamamatsu R3600 20-inch PMTs to provide both an active muon veto and passive shielding against these backgrounds. This is the first water Cherenkov shield used in a dark matter search experiment. To perform energy and position reconstruction calibrations a portal has been prepared along the central vertical axis (z-axis) of the PMT support structure through which an OFHC copper rod can be inserted directly into the target volume. Thin cylindrical calibration sources containing either of ^{55}Fe, ^{57}Co, ^{109}Cd, or ^{241}Am are placed at the tip of this rod to perform detector calibrations. A more detailed description of the XMASS detector is presented in [19].

The data used for this analysis, corresponding to 6.70 days of livetime, was taken in February 2012 with a low trigger threshold of four PMT hits [20]. Using 122 keV gammas from the ^{57}Co calibration source the xenon light yield was found to be 14.7 photoelectrons/keVee. This large light yield allows the analysis threshold to be lowered sufficiently for sensitivity to low mass WIMPs. In order to achieve optimal sensitivity, the entire detector volume is used because fiducialization is increasingly difficult at these low energies. A sequence of data reduction is applied to remove events caused by the tail of the scintillation light distribution after energetic

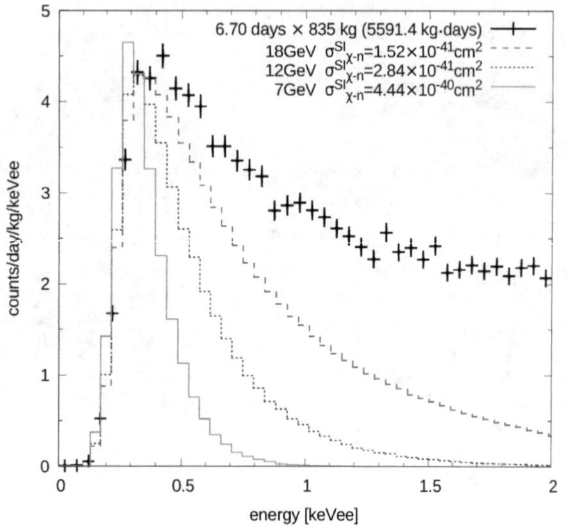

Figure 9. Simulated WIMP energy spectra in the XMASS detector assuming the maximum cross section that provides a signal rate no larger than the observation in any bin above 0.3 keVee.

events; (1) events triggered only with the liquid xenon detector are selected, (2) events that occurred within 10 ms of the previous event are rejected, and (3) events whose timing distribution has an RMS greater than 100 ns are removed. The last cut is applied to remove Cherenkov events originated from ^{40}K contamination in the PMT photocathodes; events with more than 60% of their PMT hits occurring within the first 20 ns of the event window are removed as Cherenkov-like.

Figure 9 shows simulated WIMPs energy spectra overlaid on the observed spectrum after the data reduction was applied. WIMPs are assumed to be distributed in an isothermal halo with $v_o = 220\,\mathrm{km/s}$, a galactic escape velocity of $v_{\mathrm{esc}} = 650\,\mathrm{km/s}$, and an average density of $0.3\,\mathrm{GeV/cm^3}$. In order to set a conservative upper bound on the spin-independent WIMP-nucleon cross section, the cross section is adjusted until the expected event rate in XMASS does not exceed the observed one in any energy bin above the analysis threshold. It is chosen as the energy at which the trigger efficiency is greater than 50% for 5GeV WIMPs and corresponds to 0.3 keVee. The resulting 90 % confidence level (C.L.) limit is shown in Figure 10. The impact of the uncertainty from $\mathcal{L}_{\mathrm{eff}}$ [21] is large in this analysis, so its effect on the limit is shown separately in the figure.

After careful study of the events surviving the analysis cuts, their origins are not completely understood. Contamination of ^{14}C in the GORE-TEX® sheets between the PMTs and the support structure may explain a fraction of the events. Light leaks through this material are also suspect. Nonetheless, the possible existence of a WIMP signal hidden under these and other backgrounds cannot be excluded. Although no discrimination has been made between nuclear-recoil and electronic events, and many events remain in the analysis sample, the present result excludes part of the parameter space favored by other measurements [13, 14, 15] when those data are interpreted as a signal for light mass WIMPs.

The fiducial volume analysis is ongiong for more sensitive searches. In addition, we are working on modifications to the inner surface of XMASS, especially around the PMTs, to improve the detector performance. Finally, a R&D project aiming at modification to single phase liquid xenon TPC with charge readout has been started.

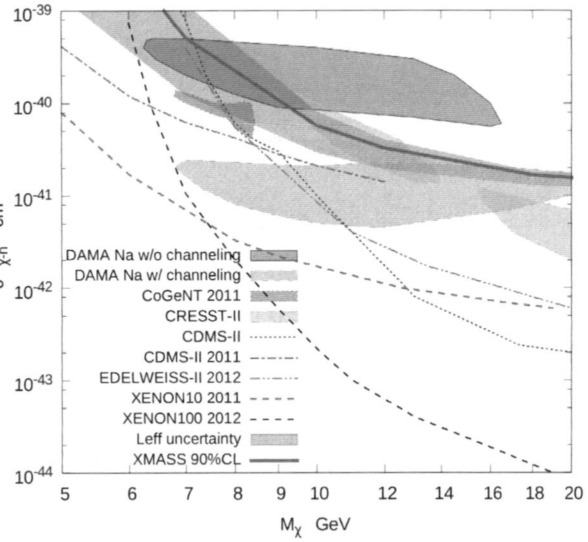

Figure 10. Spin-independent elastic WIMP-nucleon cross section as a function of WIMP mass. All systematic uncertainties except that from $\mathcal{L}_{\rm eff}$ are taken into account in the XMASS 90 % C.L. limit line. The effect of the $\mathcal{L}_{\rm eff}$ uncertainty on the limit is shown in the band. Limits from other experiments and favored regions are also shown [13, 14, 15, 16, 17, 18, 22].

References

[1] J. Hosaka et al. 2006 *Phys. Rev. D* **73** 112001.
[2] J. P. Cravens et al. 2008 *Phys. Rev. D* **78** 032002.
[3] K. Abe at al. 2011 *Phys. Rev. D* **83** 052010.
[4] B. Aharmim et al. arXiv:1109.0763.
[5] B. T. Cleveland et al. 1998 *Astrophys. J.* **496** 505.
[6] J. N. Abdurashitov et al. 2009 *Phys. Rev. C* **80** 015807.
[7] G. Bellini et al. 2011 *Phys. Rev. Lett.* **107** 141302.
[8] A. Gando et al. 2011 *Phys. Rev. D* **83** 052002.
[9] G.Jungman, M.Kamionkowski, and K.Griest, Phys. Rep. 267 (1996) 195.
[10] H. Nishimura et al. 2009 *Astropart. Phys.* **31** 185.
[11] H. Sekiya et al. 2006 *Nucl. Instr. Meth. A* **573** 204.
[12] K. Miuchi et al. 2010 *Phys. Lett. B* **686** 11.
[13] R. Bernabei et al. 2008 *Eur. Phys. J. C* **56** 333; R. Bernabei et al. 2010 *Eur. Phys. J. C* **67** 39.
[14] C. E. Aalseth et al. 2011 *Phys. Rev. Lett.* **106** 131301.
[15] G.Angloher et al. 2012 *Eur. Phys. J. C* **72** 1971.
[16] E. Aprile et al. arXive:1207.5988; J. Angle et al. 2011 *Phys. Rev. Lett.* **107** 051301.
[17] Z. Ahmed et al. 1619 *Science* **327** 1619; Z. Ahmed et al. 2011 *Phys. Rev. Lett.* **106** 131302.
[18] E. Armengaud et al. 2012 *Phys. Rev. D* **86** 051701(R).
[19] K. Abe et al. arXive:1301.2815.
[20] K. Abe et al. 2013 *Phys. Lett. B* **719** 78.
[21] E. Aprile et al. 2011 *Phys. Rev. Lett.* **107** 131302.
[22] J. Angle et al. 2011 *Phys. Rev. Lett.* **107** 051301.

Journal of Physics: Conference Series **460** (2013) 012018 doi:10.1088/1742-6596/460/1/012018

The Sanford Underground Research Facility at Homestake

S Fiorucci[1], M G Gilchriese[2], and K T Lesko[2] for the Sanford Underground Research Facility

[1] Brown University, Dept. of Physics, 182 Hope St. Providence, RI 02912

[2] Lawrence Berkeley National Laboratory, 1 Cyclotron Road, Berkeley, CA 94720

fiorucci@brown.edu

Abstract. The Sanford Underground Research Facility (SURF) at Homestake is presented. The Davis campus is described in detail including the two laboratory modules at the 4850-foot level (>4200 mwe). These modules house the LUX dark matter and MAJORANA DEMONSTRATOR neutrinoless double-beta decay experiments. The Long Baseline Neutrino Experiment plans to place their far detector at SURF. The facility is managed for the US Department of Energy (DOE) by Lawrence Berkeley National Laboratory. The South Dakota Science and Technology Authority (SDSTA) owns and operates the facility. SURF is a dedicated science facility with significant expansion capability.

1. Introduction

The initial concepts for SURF were developed with the support of the US National Science Foundation (NSF) as the primary site for the NSF's Deep Underground Science and Engineering Laboratory (DUSEL) [1]. With the National Science Board's decision to halt development of a NSF-supported underground laboratory, the US Department of Energy (DOE) now supports the operation of the facility. Both the NSF and DOE support experiments at SURF.

SURF is being developed in the former Homestake Gold Mine, in Lead, South Dakota. Barrick Gold Corporation donated the site to the State of South Dakota in 2003, following over 125 years of mining. Mining operations created over 600 km of tunnels and shafts in the facility, extending from the surface to over 8000 feet below ground. The mining levels are distributed ~150 feet apart and are referenced by the feet below the entrance to the facility.

The South Dakota philanthropist, T. Denny Sanford, gifted US$70M, to convert the former mine into a research laboratory and develop a science education facility. With these funds and the State of South Dakota appropriations, access to the underground has been reestablished and the primary access rehabilitated and improved. The facility has been stabilized and the accumulated underground water has been pumped below the 6000L. The Davis Cavity at the 4850L has been enlarged and adapted primarily for current and next-generation dark matter experiments. A new laboratory has been excavated and outfitted adjacent to the Davis Cavity to host a neutrinoless double beta decay experiment. Additional science efforts are hosted throughout the facility, including an ultrapure detector development laboratory, multiple geophysics and geological efforts, and a public outreach program.

The science program for the coming ~ five years consists of the MAJORANA DEMONSTRATOR experiment, the Large Underground Xenon (LUX) dark matter search, the Center for Ultralow Background Experiments at DUSEL (CUBED), and geoscience installations. Plans are being developed to host the Department of Energy's Long-Baseline Neutrino Experiment (LBNE), a nuclear astrophysics program involving underground particle accelerators, and second and third generation dark matter experiments.

2. Facility Description

SURF consists of surface and underground campuses and supporting infrastructure at the Homestake site. The Ross Surface Campus will be used primarily for construction and operations support, while the Yates Surface Campus will support science and administrative activities, education and public outreach functions, and the Waste Water Treatment Plant. The Ross and Yates shafts provide redundant underground access.

2.1. Yates Campus

The scientific users and facility administration are supported using three repurposed buildings: The Administration building, the Yates Education and Outreach building, the LUX warehouse and the Waste Water Treatment Plant. The Administration building provides space to support the facility administration, the scientific and facility staff, conference rooms and IT. The Yates E&O building supports the active outreach program and provides additional meeting space. The LUX Surface Laboratory provides an assembly and prototyping space for experiments. Additional buildings provide for receiving, cold storage, and the rock core archive. Scientific access to the Davis campus is primarily provided through the Yates shaft. Utilities including power, IT, ventilation and water are routed down the Yates shaft. The Yates shaft has recently been upgraded to primary access. The shaft is maintained regularly and provides 24/7 access to the 4850L and intervening levels.

2.2. Ross Campus

This campus supports facility maintenance and construction activities including skipping rock, pumping water, and general facility maintenance. Utilities including power, IT, ventilation and water are routed down the Ross shaft. The Ross shaft is in the process of undergoing a major rehabilitation replacing all the shaft furnishing, updating ground control and utilities.

2.2.1. Davis Campus

Experiments are housed in the Davis Cavity (9m x 10m x 17m) and the Transition Area (15m x 5m x 41m). Three experiments will move into the Davis Campus in 2012/2013. The LUX and CUBED experiments share the Davis Cavity. The MAJORANA experiment in the Transition Area maintains a minimum of class 10,000 cleanliness standards throughout the experimental space, including a dedicated machine shop and future electroforming room. A class 1000 clean room and glove box provide even higher standards of cleanliness to prevent contamination of the ultra- pure copper and germanium used in this experiment.

Services provided within the Davis Campus include fire sprinklers and alarms throughout the area, potable and non-potable (industrial) water, lighting, emergency lighting, ventilation, and air conditioning. A building management system provides controls throughout the Campus. No heating is required at this level due to natural rock temperatures, auto-compression of air as it travels through the shaft, and heat generated from the services at the level. Cooling is provided with two redundant 50-ton (633 M-Joule) chillers supplying chilled water to three air handling units that provide ventilation to separate Campus spaces. Chilled water is also available for experiments to connect equipment directly.

A dedicated 1500 kVA substation provides sufficient capacity for the experiment and facility needs, with margin for future expansion. Emergency power for lighting is provided with batteries in the lighting system to provide immediate light, while a standby diesel generator near the campus provides up to 24 hours of power to all safety systems in the campus. This includes water pumps in the nearby Yates shaft to prevent water from rising into the campus spaces.

Occupancy in the Davis Campus is presently limited by applicable codes to a total of 48 persons: 25 in the Davis Cavity, and 23 in the Transition Space. Shared restrooms are provided in the transition space for both campuses, as well as shower facilities for access to the clean spaces. A "cart wash" area provides the ability to clean items beyond the less clean general spaces, but prior to entry into the laboratories.

Doors and walls are installed to provide at least two isolated means of egress from any space in the event of a fire or other emergency. A system of sumps and pumps provide the ability to store grey water (not sewer water) and allow testing to ensure environmental compliance prior to pumping to the facility dewatering system. A dedicated compressor provides 100 cfm of compressed air capacity for general use.

Existing and planned experiments in the Davis Campus use liquid nitrogen to provide cooling and/or radon purging. The two primary spaces each have their own alcoves for the storage of these cryogens. The LUX experiment uses liquid xenon as the target material, requiring dedicated storage capacity, as well as an emergency storage reservoir. LUX also uses an 8 m diameter water shield to shield residual neutron and ionization radiation from the surroundings. The water for the shield is purified underground with a dedicated reverse osmosis system.

The SURF's goal has always been to provide the lowest possible radioactivity environment for experiments hosted within the facility. This commitment had been integrated into the site preparation process from the early days of the facility design, and carried over to the realization of the 4850 Level Davis Campus laboratories. We have drawn extensively from our experiences gained from SNOLab, Gran Sasso, Kamioka, and other underground research facilities to optimize the environment for the experiments. These efforts are summarized in a recent article [2]. These efforts include site and environmental characterization including rock radioactive measurements, use of low radioactivity construction materials, and regular monitoring of environmental factors including air-borne radon. Table I presents some of the assay results, obtained by direct gamma counting for rock samples from the mine, including those collected on and close to the 4850L [3].

It was found that the U/Th/K radioactivity for the underground bedrocks at Homestake is in general very low; some samples are in the sub-ppm levels. However, samples from Rhyolite intrusions, *a very small fraction of the total*, show a relatively high content of U, Th, and K more typical of the levels found in other laboratories, in particular those in granitic formations. Regions of potential Rhyolite intrusions have been identified and documented. In some cases local shielding significantly mitigates the impact of the Rhyolite intrusions.

The use of shotcrete for ground stabilization is a common practice for long-term occupancy of underground facilities; however low-radioactivity shotcrete is costly to create. The introduced construction materials have also been assayed and typical results are presented in Table 1. We have documented the distribution and thickness of the shotcrete using laser scans prior to and post application of shotcrete to provide input for detailed experimental models.

Table I. Partial U/Th/K assay results for Homestake rock samples. Overall errors estimated to be ~10-20%. Also shown are results for various construction materials (shotcrete/concrete).

	Uranium (ppm) Ave. [Range]	Thorium (ppm) Ave. [Range]	Potassium (%) Ave. [Range]
U/G Country Rock	0.22 [0.06-0.77]	0.33 [0.24-1.59]	0.96 [0.10-1.94]
Shotcrete	1.89 [1.74-2.23]	2.85 [2.00-3.46]	0.88 [0.41-1.27]
Concrete Blocks	2.16 [2.14-2.18]	3.20 [3.08-3.32]	1.23 [1.27-1.19]
Rhyolite Dike	8.75 [8.00-10.90]	10.86 [8.60-12.20]	4.17 [1.69-6.86]

In addition to the direct gamma counting of the rock and construction materials, measurements of airborne radon are ongoing and *in situ* measurements of gamma rays at the 4200 mwe depth, neutrons and cosmic ray muons have been published [4].

3. Proposed Experiments

3.1. Long-Baseline Neutrino Experiment (LBNE)

The DOE is developing plans to house a ~10 kt liquid argon detector at SURF to conduct searches for CP violation physics coupling Fermilab's powerful neutrino beams to a far detector at Homestake; comprehensive and precision measurements of neutrino mixing and oscillations; and a wide spectrum of non-beam related research including proton decay, and astronomical neutrino observations. Both surface and underground options are being developed to comply with DOE's request to establish a phased approach to the LBNE program. The surface option would be located in the Kirk Canyon to the southeast of the Ross campus. Underground options are located on the 4850L. LBNE is further described in [5].

3.2. Generation-2 Dark Matter Experiments

The DOE and NSF have plans to develop Generation-2 Dark Matter experiments, with masses ~ several tonnes. The LUX-ZEPLIN (LZ) Collaboration has submitted a proposal to develop a G-2 experiment in SURF using the existing infrastructure in the Davis Campus. SURF is open to working with all interested collaborations to develop proposals for G-2 experiments [6].

3.3. Generation-3 Dark Matter Experiments

The DOE has discussed developing a "roadmap" for Dark Matter searches including Generation-2 and Generation-3 Dark Matter experiments beginning ~ FY17. Initial efforts to develop the Generation-3 collaboration are advancing [7].

3.4. Dual Ion Accelerators for Nuclear Astrophysics (DIANA)

As part of the NSF support for the development of underground physics experiments {http://www.nsf.gov/pubs/2012/nsf12043/nsf12043.pdf} a proposal has been developed to create an underground accelerator facility for low energy nuclear astrophysics experiments. Existing excavations at the 3950L and 4850L within SURF has been identified to potentially house DIANA [8].

4. Summary

SURF is a deep underground research facility, dedicated to scientific uses and is not compromised by competing uses such as mining, transportation, or tourism. The Davis campus was recently completed and the LUX and MJD experiments are being operated at the 4850L in state-of-the-art facilities. The site benefits from the naturally low background rock and effective efforts during construction to control the introduction of trace amounts of radioactivity. The transition of the support of the facility operation from NSF to the DOE is well advanced. Stable operations of the facility were achieved in FY12, including a variety of facility improvement programs further customizing the infrastructure. SURF welcomes collaborations seeking deep underground facilities to host additional instruments and experimental efforts.

References

[1] Lesko, K.T., et al., "Deep Underground Science and Engineering Laboratory - Preliminary Design Report", arXiv:1108.0959, (2012)
[2] Chan, Y.D., "The Low-Background Construction of Laboratories at the 4850-ft Level Davis

Campus", http://dusel.org/html/early-science-progress.html, (2012)More references

[3] W. Roggenthen and A.R. Smith, White Paper: U, Th, K contents of materials associated with the Homestake DUSEL site, Lead, South Dakota. (Private Communication);
A.R. Smith, Homestake Samples: Results of Radiometric Analyses at LBNL (Private Communication)

[4] Barker, D. and Mei, D.M., "Muon-Induced Background Study for an Argon-Based Long Baseline Neutrino Experiment", Nucl.Instrum.Meth.A638:63-66,2011 arXiv:1202.5000;
Zhang C., Thomas, K., Gray, F., "Early Results on Radioactive Background Characterization for Sanford Laboratory and DUSEL Experiments", arXiv:0912.0211, (2012) BAPS.2007.DNP.DA.78, and BAPS.2010.DNP.HD.5

[5] http://lbne.fnal.gov

[6] http://science.doe.gov/grants/pdf/SC_FOA_0000597.pdf, http://www.nsf.gov/pubs/2012/nsf12043/nsf12043.pdf

[7] Salamon, M., "Second Generation Dark Matter Experiment Program", http://science.energy.gov/~/media/hep/hepap/pdf/march2012/Salamon_HEPAP_Talk_G2_DM_March_2012.pdf (2012);
Sadoulet, B., "DUSEL and the US Dark Matter program", http://www.physics.ucla.edu/hep/dm10/talks/sadoulet1.pdf (2010)

[8] Dual Ion Accelerators for Nuclear Astrophysics", http://www.jinaweb.org/underground/DIANA/

Sixth Symposium on Large TPCs for Low Energy Rare Event Detection IOP Publishing
Journal of Physics: Conference Series **460** (2013) 012019 doi:10.1088/1742-6596/460/1/012019

GridPix application to dual phase TPC

Matteo Alfonsi[1], Niels van Bakel, Auke-Pieter Colijn, Michael Patrick Decowski, Harry van der Graaf, Rolf Schön, Andrea Tiseni
Nikhef, PO box 41882, 1009 DB Amsterdam, The Netherlands

E-mail: malfonsi@nikhef.nl

Abstract. GridPix is a gas-filled detector with an aluminium mesh stretched 50 µm above the Timepix CMOS pixel chip. This defines a high electric field where gas amplification occurs. A feasibility study is ongoing at Nikhef for the application of the GridPix technology as a charge sensitive device in a dual phase noble gas Time Projection Chamber (TPC), within the framework of the DARWIN design study for next generation dark matter experiments. The smallness of the device and well defined materials allow for high radio-purity and low outgassing. The high granularity of a pixel readout and the high detection efficiency of single electrons of GridPix bring benefits especially in terms of energy resolution for small energy deposits. This feature is interesting also for the measurement of the scintillation yield and the ionisation yield of noble liquids. The accurate measurements of such quantities have a direct impact on the data interpretation of dark matter experiments. The application in dual phase argon or xenon TPCs implies several technological challenges, such as the survival of the device at cryogenic temperature as well as the operation in a pure noble gas atmosphere without discharges. We describe here the recent developments of the project.

1. Introduction

GridPix [1] is a gas-filled detector where an aluminium mesh, the *grid*, is stretched 50 µm above the pixelated Timepix CMOS chip. It is realised by means of wafer post processing technology. The Timepix chip is characterised by a matrix of 256×256 pixels and a pitch of 55 µm. The picture of the microscopic structure in Fig. 1 shows the *pillars*, made of an insulating material[2] and supporting the grid, as well as the pixels pattern on the chip surface. A thin resistive layer[3] over the chip surface protects the pixel circuitry in the event of sparks in the gas.

We investigate the application of GridPix as a charge sensitive device in a dual phase Time Projection Chamber (TPC). The development is part of the alternative charge readout working package of the DARWIN project [2], a design study for the next generation dark matter search experiments with dual phase argon or xenon TPCs.

These detectors are already playing a leading role in dark matter searches, as demonstrated by the latest results of XENON100 experiment [3]. The key concept is the simultaneous measurement of the ionisation and the scintillation light produced by a particle interaction in the liquid phase, which is the sensitive volume of the apparatus. The principle of operation is depicted in Fig. 2. The prompt scintillation light, the so-called *S1* signal, is detected by

[1] Corresponding author
[2] *SU-8 photo-resist* at the current stage [4] [5]
[3] *Si-enriched silicon nitride* at the current stage [5]

Journal of Physics: Conference Series **460** (2013) 012019 doi:10.1088/1742-6596/460/1/012019

photodetectors arrays, typically placed at the bottom of the liquid volume and in the vapour phase above the liquid. The electrons from ionisation are drifted to the liquid surface and extracted to the vapour phase by the applied electric field. Electrons, subject to a higher electric field in this region, produce *electroluminescence* as they collide with the gas atoms along the path to the anode electrode. This delayed scintillation light, the so-called *S2* signal, is detected again by the photodetectors arrays.

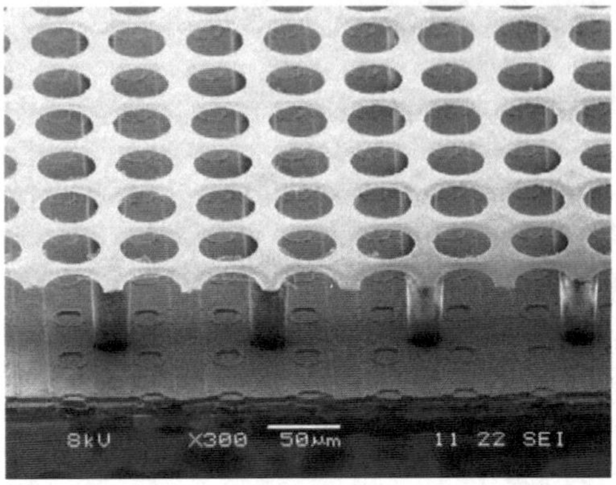

Figure 1. The microscopic structure of a GridPix device.

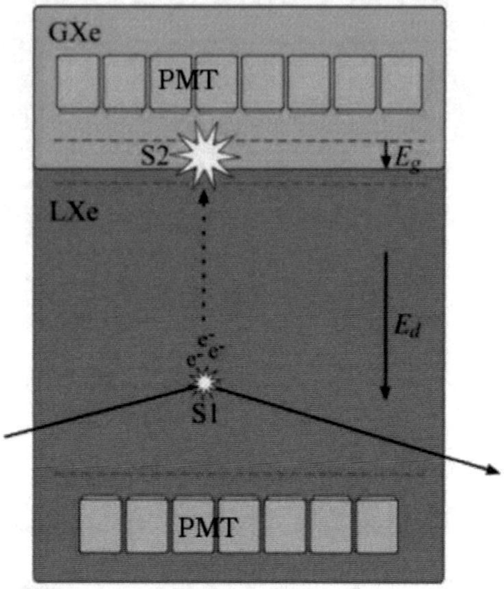

Figure 2. The typical scheme of a dual phase xenon (or argon) Time Projection Chamber.

The delay between the *S1* and the *S2* signals gives the Z-coordinate information, leading to a tri-dimensional position reconstruction. The ratio between the light and ionisation production depends on the density of the energy deposition and can be exploited for particle identification. *Electronic recoils*, i.e. electromagnetic interactions that transfers the energy to an atomic electron, are characterised by a higher charge to light ratio with respect to *nuclear recoils*, produced by neutral particles that transfer the energy to a nucleus.

2. GridPix as direct charge readout

An array of GridPix devices can instrument the TPC in place of the top photodetectors array. In this case the electroluminescence method is not required, but the electrons are collected through the grid holes in the GridPix amplification gap.

The high granularity of a pixel readout can improve the spatial resolution and the discrimination of double scatters. However it is especially effective in order to improve the energy resolution of the detector. The pixels' low noise and the high efficiency in detecting single electrons (close to 100% in Ar/iC_4H_{10} gas mixtures) allow for a digital readout approach. The electron cloud size becomes broader along the drift (in particular in the gas phase) because of diffusion and each electron is very likely collected in a different GridPix hole and detected by a different pixel: therefore the number of ionisation electrons can be inferred from the number of hit pixels.

These features are also interesting for fundamental measurements of the noble liquid scintillation and ionisation yields. The accurate knowledge of such quantities is crucial for dark matter search experiments that adopt noble liquids as target. As an example, the deposited energy is inferred from the amount of scintillation light or ionisation electrons produced. The measurements are performed in dedicated small size setups. As depicted in Fig. 3a, Compton scattering or neutron elastic scattering is exploited in order to deduce the energy deposition from the energy and the direction of the scattered photon or neutron, respectively. Compton scattering is used to obtain the scintillation and ionisation production as a function of the deposited energy for electronic recoils, while neutron scattering allows to understand the response of the medium to nuclear recoils. Systematic errors in these measurements are coming from the uncertainty on the exact spot of the interaction within the active volume (Fig. 3b) and double scatters (Fig. 3c). GridPix can provide an accurate position reconstruction reducing such systematics.

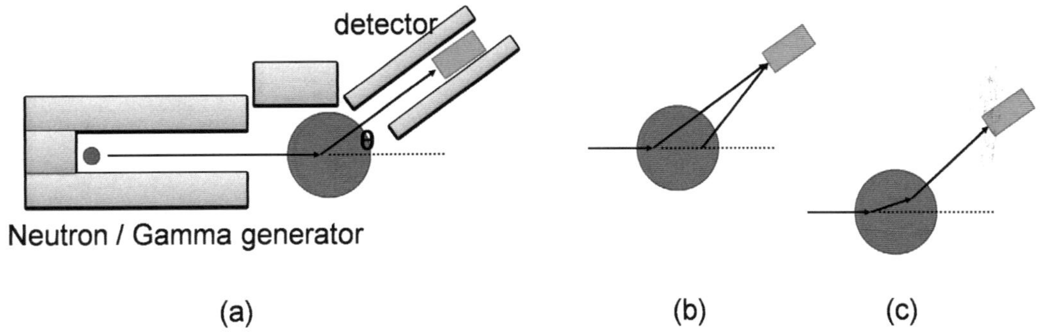

Figure 3. Scheme of the measurements of the scintillation and the ionisation yields of noble liquids for electronic and nuclear recoils. (a) Small size noble liquid detectors (blue) are irradiated by a gamma source (for electronic recoils) or a neutron source (for nuclear recoils). Compton scatterings or neutron elastic scatterings are selected by an additional detector at a specific angle θ. The energy deposited in the noble liquid is deduced from the energy and the kinematics measured with the additional detector. Sources of systematic errors are the uncertainty of the interaction position within the liquid target (b) and double scatters (c).

High radiopurity of all the components of a rare event search experiment is imperative to reduce the background. GridPix is a small device and the mass of the material introduced in the apparatus is therefore small. It is composed mainly by silicon, characterised by very low natural radioactivity. Similar arguments are valid for the outgassing properties of GridPix. The outgassing of materials is also an important property for dual phase TPC because electronegative impurities that dilute in the noble liquid capture the electrons during their drift path, hence reducing the *S2* signal.

3. GridPix operation at low temperature and development of optimised devices

The first measurements of GridPix operated in gaseous argon environment at room and low temperature as well as the test in a dual phase argon TPC are described in [7]. The test in dual phase argon revealed a mechanical weakness of the structure at cold temperature (87 K), due to the different thermal expansion coefficients of materials. As described in [8], a systematic study was performed in a temperature-controlled environment, continuously purged with nitrogen to avoid ice formation from moist air: the problem was identified in the walls, the so-called *dykes*, that support the *grid* along the perimeter of the device. Each side of the perimeter is divided in three sections. The thermal stress produces a loss of adhesion of a section from the underlying protection layer, which is then ripped off by the tension of the shrinking *grid*, as shown in Fig. 4.

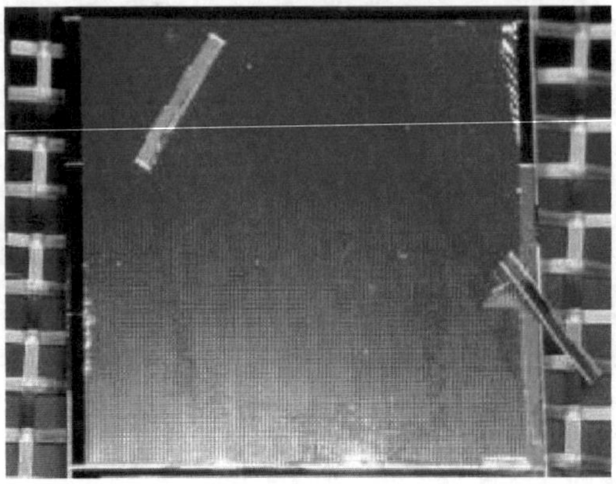

Figure 4. The *dykes* that support the *grid* along the perimeter are the weak point of the structure at cold temperature ($\sim$ 140 K in this test). Under thermal stress they lose adhesion to the underlying protection layer and the tension of the *grid* rips them off.

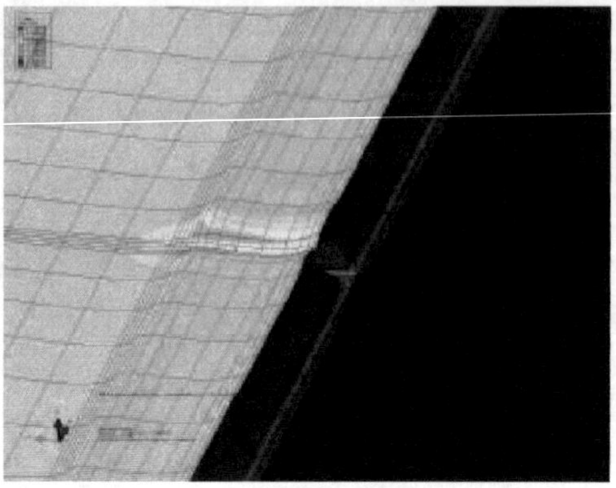

Figure 5. Engineering calculation (Finite Elements Method) of the thermal stress on the GridPix elements at a temperature of 87 K. The maximum stress is found (red region) on the edges of the sections in which the *dykes* are divided, in correspondence of the discontinuity between two sections. In this simulation the *grid* and the pillars are not included.

A similar indication is obtained by engineering calculations (using Finite Elements Methods) of the thermal stress. In Fig. 5 this is shown for a simplified structure without pillars and grid. The red regions emphasise that the discontinuity between two sections is subject to the largest stress.

We investigated with the same engineering tools modifications of the geometry that lead to more robust structures. As shown in Fig. 6, possible solutions are a finer segmentation of the dykes and the introduction of extended structures in addition to pillars for the grid support. Several prototypes with different combinations of these geometries are under production on *dummy* wafers, i.e. on a bare conductive substrate instead of a Timepix wafer. We will study the robustness at cold temperature of these new geometries. The same devices will allow to study the operation in noble gases and to compare with the standard device properties as the maximum achievable gain and gain uniformity.

At the same time, an R&D on the manufacturing process is ongoing for the realisation of a "fully ceramic GridPix", i.e. a device where pillars and *dykes* are made of silicon oxide (an insulator material) or a high resistivity silicon nitride. A successful development in this direction is a breakthrough for the low temperature operation because all the materials of the device can have matching thermal expansion properties. Low outgassing properties and high radiopurity are also expected. Furthermore the realisation of a resistive *grid* made of *Si-enriched silicon nitride* with an internal conductive aluminium network can strongly suppress the discharge development.

4. Future prospects

The GridPix production recently moved to 8" inch wafers. This important achievement demonstrates the possibility of an industrialisation and therefore the reduction of the costs.

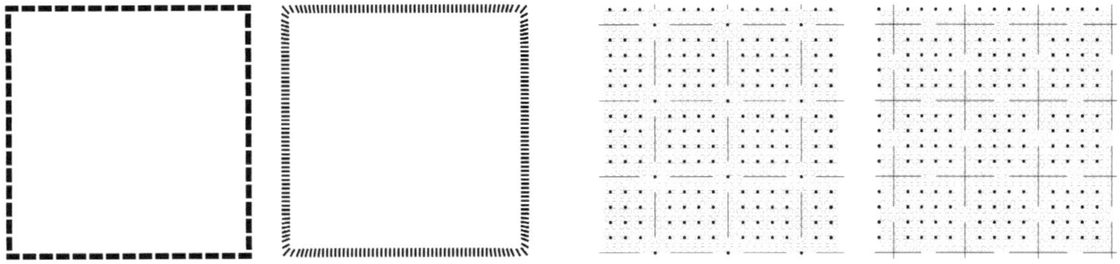

Figure 6. New geometries under study for the support of the grid that combine a finer segmentation of the grid (examples on the left) with the introduction of extended structures in addition to pillars (examples on the right).

The maximum achievable gain and the detection efficiency of single electrons will be measured with these new devices. A pressurised vessel (Fig. 7) allows to simulate at room temperature the higher gas density of a cold environment with a higher pressure. The study will be extended to pure xenon and Xe/CH_4 gas mixtures.

Furthermore we are building at Nikhef a dual phase xenon TPC (Fig. 8). The apparatus will be equipped with a gas recirculation system with hot getters to purify the xenon. We will test the GridPix operation in dual phase xenon and eventually the performance of the device as direct charge readout.

Figure 7. Setup for the measurement of the GridPix properties such as the gain and the detection efficiency of single electrons with different gas mixtures and as a function of the gas pressure.

Figure 8. Cryostat of the dual phase xenon TPC under construction at Nikhef. The detector will be built in the bottom vessel while the top vessels will house the cooling system and the heat exchanger of the gas recirculation.

References

[1] M. Campbell *et al.*, Nucl. Instr. and Meth. **A 540** (2005) 295.
[2] L. Baudis on behalf of DARWIN Consortium, "DARWIN: dark matter WIMP search with noble liquids", arXiv:1012.4767.
[3] E. Aprile *et al.*, Phys. Rev. Lett. **109** (2012) 181301
[4] M. Chefdeville, "Development of Micromegas-like gaseous detectors using a pixel readout chip as collecting anode", PhD thesis (2009), Nikhef.
[5] V.M. Blanco Carballo, "Radiation Imaging Detectors Made by Wafer Post-processing of CMOS Chips" , PhD thesis (2009), Nikhef.
[6] Y. Bilevych *et al.*, 2009 IEEE Nuclear Science Symposium Conference Records (2009) 231.
[7] M. Alfonsi *et al.*, "Development of the GridPix detector for dual phase noble gas Time Projection Chambers", 2011 IEEE Nuclear Science Symposium and Medical Imaging Conference Records (2011) 92.
[8] R. Schön *et al.*, "Operating the GridPix detector in dark matter search experiments", Nucl. Instr. Meth. A, article in press, doi 10.1016/j.nima.2012.11.054.

Liquid Hole-Multipliers: A potential concept for large single-phase noble-liquid TPCs of rare events

Amos Breskin

Department of Astrophysics and Particle Physics
Weizmann Institute of Science
76100 Rehovot, Israel
Amos.breskin@weizmann.ac.il

Abstract. A novel concept is proposed for large-volume single-phase noble-liquid TPC detectors for rare events. Both radiation-induced scintillation-light and ionization-charge are detected by Liquid Hole-Multipliers (LHM), immersed in the noble liquid. The latter may consist of cascaded Gas Electron Multipliers (GEM), Thick Gas Electron Multiplier (THGEM) electrodes or others, coated with CsI UV-photocathodes. Electrons, photo-induced on CsI by primary scintillation in the noble liquid, and event-correlated drifting ionization electrons are amplified in the cascaded elements primarily through electroluminescence, and possibly through additional moderate avalanche, occurring within the holes. The resulting charge-signals or light-pulses are recorded on anode pads or with photosensors – e.g. gaseous photomultipliers (GPM), respectively. Potential affordable solutions are proposed for multi-ton dark-matter detectors; open questions are formulated for validating this dream.

As a result of a dream, eventually based on solid experimental facts, we propose and discuss a new possible concept of a *single-phase* noble-liquid Time Projection Chamber (TPC) for large-volume detectors of rare events.

Noble-liquid detectors have been employed since a few decades in particle-physics calorimetry, gamma imaging in astronomy and medical fields (e.g. PET, Compton Camera), as well as in astroparticle physics. In the latter, most important are neutrino physics and dark-matter (DM) searches. Detailed surveys of present techniques and their applications can be found in recent reviews [1, 2, 3].

One of the most advanced methods of direct detection of DM signatures, in form of Weakly Ionizing Massive Particles (WIMPs), is the observation of nuclear-recoil rates they deposit in rare scattering events in a noble-liquid TPC detector. The extremely low interaction cross sections involved require large-mass target-detectors, with the main challenges being the detection of faint low-energy (keV-scale) signals and effective background suppression. Liquid argon (LAr, e.g. ArDM [4]) and liquid xenon (LXe, e.g. XENON100 [5], LUX [6]) are the preferred targets, allowing for the construction of large volumes of homogeneous detection media. They operate either in a *single-phase* (liquid), or *dual-phase* (liquid and gas) configurations. *Single-phase* detectors (e.g. XMASS LXe experiment [7]) are simpler, relying on measuring the scintillation light emitted promptly at the site of interaction. *Dual-phase* detectors operate as TPCs, recording two signals: the primary prompt scintillation light (S1) within the liquid and a secondary delayed signal (S2) generated by the recoil-induced ionization electrons liberated at the interaction site, as they pass through the gas phase after extraction from the liquid. S2 can be a result of electroluminescence in the saturated vapor phase above the liquid [5] or of a multiplied charge [4] in this gas gap. Background suppression is achieved by running the experiments in underground facilities, using radio-pure detector materials and applying passive shielding and active veto schemes. The distinct difference of the secondary-to-primary signals ratio (S2/S1), in *dual-phase* detectors, for nuclear recoils and for electron recoils from gamma background is the key to their efficient discrimination as demonstrated in [5].

The sensitivity of DM detectors to detect low-energy WIMP-induced recoils and effectively discriminate them from background depends largely on the photon detectors. Present-day DM detectors and others under conception (e.g. XENON1ton [8]) employ large and very costly vacuum-photomultiplier (PMT) arrays; these must withstand cryogenic conditions and have low natural radioactivity and high single-photon detection efficiency at the relevant UV-emission wavelengths. While present PMTs reasonably fulfill the strict requirements of current experiments, novel affordable solutions are required for future generations of multi-ton detectors, e.g. DARWIN [9]. For such experiments, the price of present-type photon detectors would become exorbitant – calling for new solutions.

One concept, under advanced R&D, relies on photon recording in *dual-phase* TPCs with large-area gas-avalanche photomultipliers (GPMs) [10, 11, 12]. These are expected to have lower cost, flexible experiment-adapted flat geometry, large size, high pixilation and potentially low background. A typical GPM consists of cascaded hole avalanche multipliers (e.g. Gas Electron Multipliers (GEM[13]), Thick Gas Electron Multipliers (THGEM [14])) or hybrid combinations of hole and mesh multipliers [15, 16], with the first element being coated with a robust CsI UV-photocathode [17]. Both the THGEM-GPM and the hybrid-GPM showed high gains in combination with a LXe-TPC [15, 16]. The GPM concept has been under intense R&D at the Weizmann Institute, within DARWIN [9], for future multi-ton DM detectors; it has been considered by the PANDA projected DM experiment, also in a *single-phase* configuration [18]. A *single-phase* LXe detector coupled to a GPM is under advanced R&D, for combined fast-neutron and gamma imaging [19].

While *dual-phase* noble-liquid detectors have become rather "standard" instruments in present experiments, the expansion of their geometrical dimensions in multi-ton devices might not be technically simple; particularly because of the necessity to extract electrons from liquid into gas through large, very flat mesh-electrodes, keeping constant temperature (and pressure) conditions across the detector. We therefore propose a novel challenging concept that would permit the efficient recording of both low-energy recoil-induced scintillation-light and ionization-electron signals in large-volume *single-phase* noble-liquid detectors.

In our new concept, both radiation-deposited scintillation light and ionization charges are collected within the liquid into novel Liquid Hole-Multipliers (LHM), immersed within the liquid. The LHMs consist of a combination of cascaded GEMs, THGEMs, Micro-Hole & Strip Plates (MHSPs) [20] or other dedicated electrodes, coated with CsI photocathodes; the latter have high quantum efficiency values for both LAr and LXe emission wavelengths [16], also in liquid xenon [21, 22]. The process is schematically shown in Figure 1. Electrons, photo-induced on CsI by primary scintillation and event-correlated drifting ionization electrons are collected within the liquid, by a strong dipole field, into the multiplier's amplification holes. They generate electroluminescence within the holes under high electric fields at their center, and, according to the field-strength, they may also undergo some modest charge multiplication. Forward-emitted UV-photons (shown in Figure 1) impinge on the photocathode of the next element in the cascade, inducing additional photoelectrons that generate UV photons in its holes; the process can be repeated until the right sensitivity is reached (even without charge multiplication). According to the total amplification reached, the final charge signals can be collected on readout pads; alternatively, photon flashes from the last element can be recorded by photon detectors deployed within the liquid – preferably GPMs (Figure 1). To avoid possible photon-feedback effects generated by electroluminescence (e.g. secondary electrons from electroluminescence-photons emitted backwards) the hole-electrodes of the cascade could be staggered.

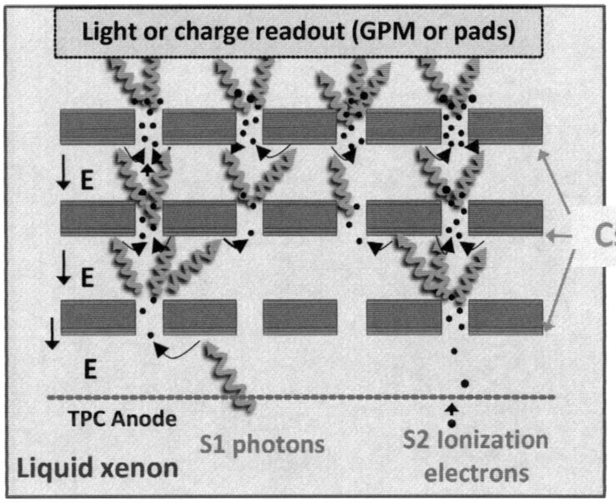

Fig. 1 The proposed Liquid Hole-Multiplier (LHM) principle. Light-amplification and optional modest charge multiplication in sensors immersed within the noble liquid permits detection of both radiation-induced scintillation UV-photons (S1) and ionization electrons (S2). Shown is an example with 3 THGEM electrodes:

- *Radiation-induced UV-photons impinge on the first CsI-coated THGEM electrode;*
- *Extracted photoelectrons are trapped into the holes, where high fields induce electroluminescence (and possibly small charge gain); The resulting forward-emitted photons (the only ones shown here) are further amplified by a cascade of CsI-coated THGEMs.*
- *Similarly, drifting radiation-induced ionization electrons are focused into the holes and follow the same amplification path.*
- *S1 and S2 signals are recorded optically by an immersed Gaseous Photomultiplier (GPM) or by charge collected on pads.*

It should be noted that the working hypotheses for this new concept have been derived from the following known facts:

- The existing knowledge that moderate charge multiplication (a few hundreds at best) [23, 24] and electroluminescence (~100 photons/electron at gain ~50) [25] are possible on a few-microns diameter anode wires immersed in LXe, as also reviewed in [26, 3]. To our best knowledge, no stable gain was reported on thin wires in LAr; charge gains of ~100 were reached only on sharp (0.25 radius) tips in LAr [27];

- A recent demonstration that electroluminescence (estimated yields reaching 500 photons/electron), without charge multiplication, occurs in THGEM holes in LAr, at rather moderate fields [28, 12];
- Similar photon-mediated amplification concepts, in a cascaded detector, were already demonstrated by our research team (Weizmann/Aveiro/Coimbra) [29] and in parallel by others [30], in a gas phase, with the aim of blocking avalanche ions.
- CsI photocathodes are known to have high quantum yields in LXe (~23% at 178nm under 10kV/cm) [21, 22], which was the basis for some potential *dual-phase* TPCs [22, 31] and *single-phase* spherical-TPC [32] noble-liquid detector concepts (not yet materialized). The latter, of a spherical geometry, was conceived with charge readout after multiplication within the liquid in a central spherical GEM detector.

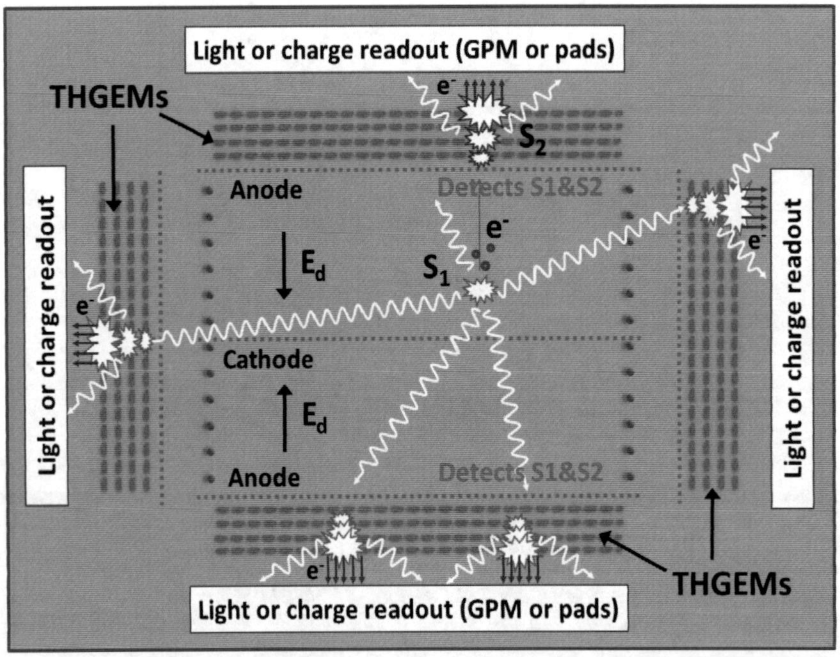

Fig. 2 *A proposed dual-sided single-phase noble-liquid TPC DM detector with top, bottom and side THGEM-LHMs (or other LHMs). The prompt S1 scintillation signals are detected with all LHMs. The delayed S2 ionization signals are recorded with the bottom and top LHMs. The S1 and S2 signals are recorded optically by an immersed GPM or by charges collected on pads.*

As the LHM is sensitive to both electrons and photons (Figure 1), one could conceive *single-phase* symmetric noble-liquid detectors, with bi-directional drift volumes, having additionally photosensors at their circumference (Figure 2). The net advantages would be: a two-fold reduced drift potential (with significant technical benefits) and a very efficient primary-scintillation signal collection over a broad solid angle – with a potentially lower detection threshold for low-mass WIMPs. A further embodiment (Figure 3) would be a large volume composed of a stack of further shorter bi-directional drift volumes with multiple-LHM elements read out by pads of flat GPMs.

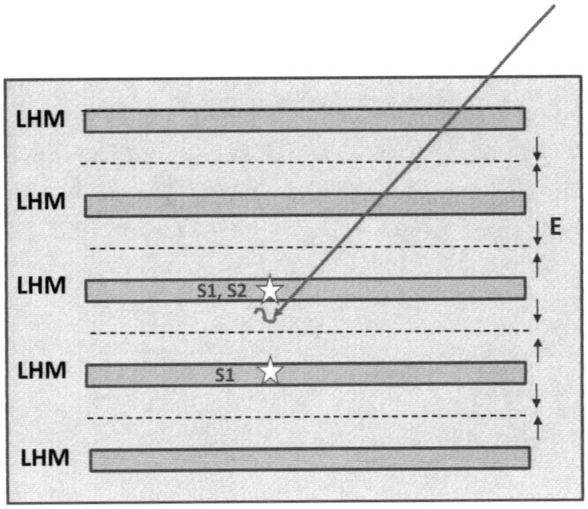

Fig. 3 *A possible layout of a large-volume single-phase noble-liquid DM detector composed of a stack of short, bi-directional, drift gaps with multiple double-sided LHM elements. The S1 and S2 signals induced by an event are recorded by pads of by flat GPMs. Drawing not to scale.*

The validation of this challenging *dream-concept* requires crossing many potential pitfalls. The studies are in course in the novel LXe cryostat system (WILiX) assembled at the Weizmann Institute; they necessitate careful investigations of numerous relevant issues and parameters. A success will rely first of all on the possibility of reaching sufficient light amplification, and maybe small charge multiplication, in liquid phase, in single- and cascaded-multipliers and validating the photon-assisted multiplication in liquid phase [29]. Unlike single-wire structures [26], cascaded elements have the potential of reaching high total gains, with a moderate gain per single element, e.g. as shown in [16]. While light amplification onset in THGEM holes was measured in LAr at relatively low fields at the hole center (~ 60kV/cm) [28, 12], some literature results [26] indicate upon higher field values, of ~100 and 1000kV/cm necessary for the respective onset of scintillation and charge multiplication on thin wires in LXe. If necessary, the hole-multipliers could be interlaced with thin amplification-wires; such hybrid multipliers would provide higher total light and charge yields due to the high reachable fields at the wire's vicinity at comparatively low potentials, as summarized in [26]. The multipliers' geometry must be optimized also for high collection efficiency of both ionization electrons and photoelectrons into the holes, e.g. similarly to [33]. Other issues would be the choice of high-purity electrode materials (at a later stage, also radio-pure ones), understanding the role of impurities in the liquid, photon-feedback suppression, stability, evaluating pads vs. optical readout etc. Finally, one should evaluate the suitability of the detector's pulse-height and time resolutions, detection thresholds and background suppression capabilities for rare-event detection.

The validation of the new proposed concept could have considerable impact on the conception of the next generation of large-mass LXe or LAr detectors of DM and other rare events. The potential highlights would be: a rather simple and probably economic construction of PMT-free detectors sensitive to both scintillation light (S1) and ionization electrons (S2), larger scintillation signals (very large solid angle) with resulting lower expected detection thresholds and shorter drift lengths (Figs 2, 3) resulting in lower drift voltages and lower electron losses. If successful, the new LHM concept could pave ways towards other fields of applications.

I would like to thank Dr. Lior Arazi and Dr. Rachel Chechik of the Weizmann Institute and Prof. Elena Aprile of Columbia University for helpful discussions and collaboration in this project. This work was supported in part by the Israel Science Foundation (Grant 477/10) and the MINERVA Foundation (Project 710827). A. Breskin is the W.P. Reuther Professor of Research in the Peaceful use of Atomic Energy.

References

[1] E. Aprile and T. Doke, Liquid xenon detectors for particle physics and astrophysics, *Rev. Mod. Phys.* **82** (2010) 2053–2097.

[2] D. Y. Akimov, Techniques and results for the direct detection of dark matter, *Nucl. Instrum. Meth. A***628** (2011) 50–58.

[3] V. Chepel and H. Araújo. Liquid noble gas detectors for low energy particle physics. *JINST*, in press.

[4] ARDM collaboration, C. Amsler, A. Badertscher, V. Boccone, A. Bueno, M. C. Carmona-Benitez, W. Creus, et al., First results on light readout from the 1-ton ArDM liquid argon detector for dark matter searches, 2010 *JINST* **5** P11003.

[5] E. Aprile, K. Arisaka, F. Arneodo, A. Askin, L. Baudis, A. Behrens, et al.; The XENON100 dark matter experiment, *Astropart. Phys*. **35** (2012) 573–590.

[6] D. S. Akerib, X. Bai, S. Bedikian, E. Bernard, A. Bernstein, A. Bolozdynya, et al., The large underground xenon (LUX) experiment, *Nucl. Instrum. Meth. A***704** (2013) 111-126.

[7] J. Liu and the XMASS collaboration, The XMASS 800 kg detector, 2012 *J. Phys.: Conf. Ser.* **375** 012022

[8] E. Aprile (for the XENON1T collaboration); The XENON1T Dark Matter Search Experiment. Proceedings of DM2012, Tenth Symposium on Sources and Detection of Dark Matter and Dark Energy in the Universe; UCLA, California, USA, February 2012. arXiv:1206.6288v1

[9] L. Baudis (for the DARWIN consortium); DARWIN dark matter WIMP search with noble liquids. 2012 *J. Phys.: Conf. Ser.* **375** 012028; arXiv:1201.2402v1

[10] R. Chechik and A. Breskin; Advances in Gaseous Photomultipliers; *Nucl. Instr. Meth. A***595** (2008) 116-127.

[11] A. Breskin, V. Peskov, M. Cortesi, R. Budnik, R. Chechik, S. Duval, D. Thers, A.E.C. Coimbra, J.M.F. dos Santos, J.A.M. Lopes, C.D.R. Azevedo and J.F.C.A. Veloso. CsI-THGEM gaseous photomultipliers for RICH and noble-liquid detectors. *Nucl. Instrum. And Meth. A* **639**(2011)117-120.

[12] A. Buzulutskov; Advances in Cryogenic Avalanche Detectors, 2012 *JINST* **7** C02025.

[13] F. Sauli, Progress with the gas electron multiplier Original Research. *Nucl. Instrum. And Meth. A***522**(2004)93-98.

[14] A. Breskin, R. Alon, M. Cortesi, R. Chechik, J. Miyamoto, V. Dangendorf, J. Maja, J.M.F. Dos Santos; A concise review on THGEM detectors; *Nucl. Instr. Meth. A***598**(2009)107-111.

[15] S. Duval, A. Breskin , R. Budnik, W.T. Chen, H. Carduner, M. Cortesi, J.P. Cussonneau , J. Donnard, J. Lamblin , P. Le Ray , E. Morteau , T. Oger , J.S. Stutzmann and D. Thers . On the operation of a Micropattern Gaseous UV-Photomultiplier in Liquid-Xenon. 2011 *JINST* **6** P04007.

[16] S. Duval, Arazi L, Breskin A, Budnik R, Chen W T, Carduner H, Coimbra A E C, Cortesi M, Kaner R, Cussonneau J P, Donnard J, Lamblin J, Lemaire O, Le Ray P, Lopes J A M, Hadi A F M, Morteau E, Oger T, dos Santos J M F, Lavina L S, Stutzmann J S, Thers D, Hybrid multi micropattern gaseous photomultiplier for detection of liquid-xenon scintillation. *Nucl. Instrum. Meth. A***695**(2012)163-167.

[17] A. Breskin, CsI UV photocathodes: history and mystery, *Nucl. Instrum. Meth. A* **371** (1996)116–136.

[18] K. Giboni, X. Ji, A. Tan, and L. Zhao, PANDA-X: A new detector for dark matter search. Presented at KEK Seminar, Tsukuba, Japan, Nov., 2011.

[19] A. Breskin, I. Israelashvili, M. Cortesi, L. Arazi, S. Shchemelinin, R. Chechik, V. Dangendorf, B. Bromberger and D. Vartsky. A novel liquid-xenon detector concept for combined fast-neutron and gamma imaging and spectroscopy. 2012 *JINST* **7** C06008.

[20] J.F.C.A. Veloso, J.M. Maia, L.F. Requicha Ferreira, J.M.F. dos Santos, A. Breskin, R. Chechik, Rui de Oliveira; Recent advances in X-ray detection with micro-hole and strip plate detector. *Nucl. Instrum. Meth. A***524**(2004)124-129.

[21] E. Aprile, A. Bolotnikov, D. Chen, R. Mukherjee, F. Xu, D.F. Anderson, V. Peskov. Performance of CsI photocathodes in liquid Xe, Kr, and Ar. *Nucl. Instrum. Meth. A* **338**(1994)328-335.

[22] E. Aprile, K.L. Giboni, S. Kamat, P. Majewski, K. Ni, B.K. Singh and M. Yamashita. Performance of Dual Phase Xe-TPC with CsI Photocathode and PMTs readout for the Scintillation Light. *IEEE ICDL* 2005, p345.

[23] S. E. Derenzo, T. S. Mast, H. Zaklad, and R. A. Muller, Electron avalanche in liquid xenon, *Phys. Rev. A* **9** (1974) 2582–2591.

[24] M. Miyajima, K. Masuda, A. Hitachi, T. Doke, T. Takahashi, S. Konno, et al., Proportional counter filled with highly purified liquid xenon, *Nucl. Instrum. Meth.* **134** (1976) 403–405.

[25] K. Masuda, S. Takasu, T. Doke, T. Takahashi, A. Nakamoto, S. Kubota, et al., A liquid xenon proportional scintillation counter, *Nucl. Instrum. Meth.* **160** (1979) 247–253.

[26] T. Doke, Recent developments of liquid xenon detectors, *Nucl. Instrum. Meth.* **196** (1982) 87–96.

[27] G. Bressi, M. Cambiaghi, G. Carugno, E. Conti, and E. D'Uscio, Electron multiplication in liquid argon on a tip array, *Nucl. Instrum. Meth. A* **310** (1991) 613–617.

[28] P. K. Lightfoot, G. J. Barker, K. Mavrokoridis, Y. A. Ramachers, and N. J. C. Spooner, Optical readout tracking detector concept using secondary scintillation from liquid argon generated by a thick gas electron multiplier, 2009 *JINST* **4** P04002.

[29] J.F.C.A. Veloso, F.D. Amaro, J.M.F. dos Santos, A. Breskin and R. Chechik; The Photon-Assisted Cascaded Electron Multiplier: a Concept for Avalanche-Ion Blocking. 2006 *JINST* **1** P08003.

[30] A Buzulutskov and A Bondar. Electric and Photoelectric Gates for ion backflow suppression in multi-GEM structures. 2006 *JINST* **1** P08006

[31] M.Gai, R.Alon, A.Breskin, M.Cortesi, D.N.McKinsey, J.Miyamoto, K.Ni, D.A.R.Rubin & T.Wongjirad; toward application of a thick gas electron multiplier (THGEM) readout for a dark matter detector. Proc. 23rd Winter Workshop on Nuclear Dynamics, Big Sky, Montana, USA, Feb. 11-18, 2007. arXiv:0706.1106

[32] P. Majewski, R&D for future ZEPLIN: a multi-ton LXe DM detector. Presented at CRYODET – Cryogenic Liquid Detectors for Future Particle Physics, LNGS, Italy, Mar, 2006.

[33] C.D.R. Azevedo, M. Cortesi, A.V. Lyashenko, A. Breskin, R. Chechik, J. Miyamoto, V. Peskov, J. Escada, J.F.C.A. Veloso, J.M.F. dos Santos. Towards THGEM UV-photon detectors for RICH: on single-photon detection efficiency in Ne/CH4 and Ne/CF4. 2010 *JINST* **5** P01002.

AUTHOR INDEX

AUTHOR INDEX